BENJAMIN HOLT & CATERPILLAR
TRACKS & COMBINES

Reynold M. Wik

American Society of Agricultural Engineers

The National Endowment for the Humanities, Washington, D. C., and The Haggin Museum, Stockton, California, have provided funds for the research and writing of this publication.

LCCN: 84-72594
ISBN: 0-916150-66-6

Printed in the United States of America by the American Society of Agricultural Engineers, 2950 Niles Road, St. Joseph, MI 49085-9659.

PREFACE

"Benjamin Holt and Caterpillar: Tracks and Combines" is the first definitive history and scholarly study on the life and inventive genius of Benjamin Holt. Not only does it portray his invention and successful development of the crawler tractor, but also the advancement he made in improving the combined harvester.

In the past, virtually all historical writing ignored the personal life of Benjamin Holt and his brothers, and failed to depict the story of the people behind the machines and the Holt Manufacturing Company, which was the beginning of what is today the Caterpillar Tractor Co. Company histories that have appeared tend to oversimplify the process involved—to the reader, the steps seem simple: invent a machine, build a factory and get rich. Dr. Wik's account is written not only to illustrate the success stories, but also to depict the hardships and difficulties involved in managing a large industry.

In the early 1900s, there were approximately four hundred companies engaged in the production of farm tractors, but today there are only a handful that remain and one of these, the Caterpillar Tractor Co. of Peoria, Illinois, had its early beginnings here in Stockton, California. It was here that the inventive genius, Benjamin Holt, had the foresight, perseverance, stamina and integrity to build machinery that has literally changed the face of the planet on which we live.

The Haggin Museum is fortunate and priviledged to exhibit early examples of this machinery, along with original scale models, a portion of Benjamin Holt's experimental shop and other material pertaining to the early accomplishments in Stockton. In addition to these representative examples of early equipment, the Museum maintains an extensive Holt Manufacturing Company and industrial history archives devoted exclusively to local contributions to agricultural and industrial technology. Just recently, the archives was awarded a matching grant from the William Knox Holt Foundation to establish a special area for this purpose and to employ an archivist to catalog and cross reference the collection, presently numbering over 60,000 items.

I would like to take this opportunity to pay tribute to the late William Knox Holt, last surviving son of Benjamin Holt, and the late Warren H. Atherton for making the original work on this publication possible and for the many hours of oral history interviews they (and others) granted, without which the realization of such a definitive work as this could not have been possible.

In addition, I thank Reynold M. Wik for his exhaustive efforts, including his application for and securing of a National Endowment for the Humanities grant which made possible the further research that was so vital for the complete story.

Keith E. Dennison, Director
The Haggin Museum

TABLE OF CONTENTS

FOREWORD

It has been a gratifying experience to write this volume about Benjamin Holt because he was an inventor and industrialist who manufactured machines which lightened the burden of labor for working people in America. His development of the combined harvester revolutionized crop harvesting, and his first successful track-type tractors represented a major engineering improvement in the use of the wheel. His achievements stand in importance with the contributions of Cyrus McCormick, John Deere and Henry Ford to the history of technology in rural America.

A number of thoughtful people have assisted in the preparation of this book. Keith E. Dennison and Raymond W. Hillman of The Haggin Museum in Stockton, California, were instrumental in initiating this project and making available important archival materials.

The members of the Executive Board of The Haggin Museum provided a research grant to assist in doing research in Concord, New Hampshire, the National Archives in Washington D.C., the Caterpillar Tractor Co. in Peoria, Illinois, and the Shields Library at the University of California in Davis. In addition, a fellowship grant from the National Endowment for the Humanities assisted in the completion of this work.

Frank H. Holt and Pliny G. Holt of Potomac, Maryland, gave me permission to use the Pliny E. Holt Papers which they had rescued from the Stockton city garbage dump, and which were later donated to The Haggin Museum in Stockton. This collection is a treasury of primary source materials relating to the period when the first track-type tractors were designed and became commercially successful.

The late William K. Holt, of Hillsborough, California, granted numerous interviews at his home and provided important material pertaining to the life of his father. Likewise, Warren H. Atherton supported the biography of his father-in-law and provided several taped interviews about his own life in Stockton.

Many people assisted my research. In Concord, New Hampshire, Jean G. Johnson at the New Hampshire Historical Society, Majorie B. Foote and Virginia B. Drivas at the city clerk's office, and Henry Dowst, Jr. of the Superior Court helped piece together the Holt family's history. At the Caterpillar Tractor Co. in Peoria, Illinois, S.M. Dabney of the legal staff; Ed Ashton, formerly of the patent department; and Fred Tuerk, of public relations, directed me to important historical materials.

Several Stockton residents generously helped reconstruct the life of Benjamin Holt and other Holt families: Constance Miller Eleanor Holt Drake, Mrs. Homer Guernsey, and Harold Noble. Harry D. Holt and Parker Holt provided important information about their father, C. Parker Holt, who was a Holt Manufacturing Company executive for over 30 years. Other former employees of this corporation who shared their recollections include Joe Reposa, Fred M. Ballew, Walter Gilgert, Jack Ross, Russell C. Grigsby, and Harold (Jack) Baker. James F. Page of Alameda, California, related details about his father, Myron, who lived with the Benjamin Holt family for many years.

Finally, my family shared in this endeavor, and my special thanks and appreciation go to my wife Helen, our son Denis, our daughter-in-law Teresa, and our grandsons Jason and Peter.

Reynold M. Wik

Chapter 1

THE BENJAMIN HOLT LEGACY

Benjamin Holt and his machines have literally changed the face of the earth.(1) As president of the Holt manufacturing enterprises in Stockton, California, from 1883 to 1920, he fashioned a career which influenced the lives of millions of people. He was largely responsible for two significant technological advances in engineering, namely the development and refinement of the combined harvester and the invention of the Caterpillar track-laying tractor. Subsequently the two machines advanced farming and earth moving practices around the globe.

Born in Loudon, New Hampshire, on January 1, 1849, Benjamin Holt worked with his father and four brothers in making wagon wheels. As the business expanded, the Holts established an office in San Francisco in 1868, and a factory in Stockton 15 years later. Eager to take advantage of the expanding agricultural economy in the Pacific Coast states, the Holt brothers began the manufacture of combined harvesters in 1886, followed by the production of huge steam traction steam engines in 1890. These behemoths were used for plowing, logging, mining, and combining grain in the immense fields of grain in the West.

In progressive fashion, the Holt company officials moved on to develop the revolutionary track-type tractor—an innovation which made an improvement on the use of the wheel, something which engineers had been attempting to do for several hundred years. The crawler tractor made it possible to apply greater motive power to farming, road building, lumbering, dam building, and military transport by enabling heavy machines to traverse sand, mud, snow and rough terrain on their own perpetual, broad tracks. Benjamin Holt and his associates were the first to manufacture a successful farm tractor which moved on a self-laying track rather than on conventional wheels. This historic achievement occurred in Stockton on November 22, 1904.(2)

During World War I, this track-laying principle was adapted for the building of military tanks by the British Government. General Ernest D. Swinton of the British Army, who had helped develop the tank, traveled to Stockton in the spring of 1918 to thank Benjamin Holt and his employees for their contribution to the development of this new weapon of war. During ceremonies in Stockton, Benjamin Holt agreed to appear on the platform only after intense urging by his friends. In 1921, a journalist in Concord, New Hampshire, stated that he knew of no man who had gained as much fame as Holt had who was as modest in accepting praise and recognition for his achievements.

Benjamin Holt was born in Loudon, New Hampshire on January 1, 1849. He died in Stockton, California, December 20, 1920.

Benjamin Holt posed for this picture in September 1894 at age 45.

Later in World War II, army tanks with churning treads charged into battle and bulldozers on tracks ripped through barbed wire barriers, clearing land for airfields, hacking roads through jungles, and tearing up pillboxes. Admiral William F. Halsey credited the Allied victory in the Pacific theater to the use of the submarine, the airplane, radar, and the tractor bulldozer.(3)

Called the "Edison of the West," Benjamin Holt lived modestly in a middle-class neighborhood and spent little time at social affairs. He enjoyed privacy, ignored politics, and never took public stands on political issues. His friends never knew his political persuasion, and he registered to vote only once—in 1896—and even then he avoided listing his political party preference at the Court House of San Joaquin County.(4) On the voter's registration questionnaire he had described himself as five feet nine inches tall with brown eyes and light brown hair.

Dedicated to his work, he rarely took a vacation. When he took trips out of town they were usually associated with his business interests. In fact, he didn't visit nearby Yosemite National Park until he was seventy years of age. His diversions seemed to be limited to driving in the country, attending his annual employees' picnic, and chatting with some of his factory workers during the lunch hour. His son, William K. Holt, recalled that "My father was so absorbed in his work that his only luxury seemed to be an occasional stop at a local watering hole for a glass of beer on his way home from a hard day's work."(5)

In 1890, at age 41, Benjamin Holt married Anna Brown, the daughter of a pioneer California family, and although their five children, Alfred, William, Anne, Edison, and Dean, were never in want, neither were they pampered. They attended the public schools in Stockton, and the sons completed university educations. All were taught the work ethic and the value of money at an early age.

During his career, Holt's gifts to charitable causes were made as anonymously as possible. When one of his factory workers suffered a prolonged illness, Holt continued to mail the man's regular paychecks. After a plant explosion killed an employee with a large family, the Holts raised some of the victim's children in their own home.(6)

Benjamin Holt consistently refused to give speeches in public. He even hated to talk on the telephone, claiming that he could not hear anything. In conversation he would often let others ramble on at length, uttering only a short, incisive comment to conclude the exchange. In this sense, he was thought by some to be a typical Yankee. Those who knew him intimately described him as reserved, reticent, taciturn, quiet, modest, generous, aloof and stern. A proper person, he elicited respect and people refrained from crossing him while exchanging views. Disdaining pretense, he often wore a comfortable, old and frayed black overcoat, despite suggestions from his wife that he wear something more befitting his position.

Although retiring in his social contacts, Benjamin enjoyed new experiences such as traveling in Europe in 1913 and taking a three-hour ride in a zeppelin over Berlin. Later at a San Joaquin County Fair where a barnstorming airplane pilot was giving rides, he eagerly donned helmet and goggles and rode in an open cockpit. His son, William K., once remarked that his father was not afraid of anything.

Throughout his life, Benjamin's mental powers were focused on mechanics, engineering and practical inventions. Iron-willed about reaching each new goal, he designed new machines where others had failed. Between 1889 and 1920, he received 47 patents on harvesters and traction engines and another 140 patents in collaboration with six other company engineers.(7) Since his mind could retain vast amounts of technical information, he could recall the exact specifications of scores of engines without recourse to written records. He usually did not bother with blueprints and seldom wrote anything down on paper. In fact, today there is only one item in the whole collection of business records of the Holt Manufacturing Company which contains his hand writing and this is a telegram ordering some spare engine parts. Some of his co-workers remember never seeing their boss with a book in his hands.(8)

Imbued with a love of machinery, Holt spent most of his working hours in his experimental shop at the factory. Here he donned old clothes and worked with the materials at hand. When tackling a tough problem, he put everything else out of his mind, often forgetting to stop for meals or listen to his associates. Once he came upon a new idea, nothing could distract him until he had determined its feasibility. On business trips, a traveling companion bought the tickets, planned the route and paid the bills, leaving Holt free to concentrate on technical matters.

Holt's intense mental concentration became a standing joke among his employees. As he worked, he had the habit of picking up rags used for wiping grease from machinery and sticking them in his pocket. When he needed a handkerchief, he would absentmindedly wipe his face with these oily scraps, prompting one visitor to ask a worker how long "that man" with the dirty face had been working in the plant. "Fifteen or twenty years," replied the foreman, who did not explain who that man was. "Is that all," the visitor replied, "I didn't think a man could get that dirty in twenty years."(9)

Another legendary example of Ben Holt's fixity of purpose occurred in 1903 when he returned to New York after a trip to Europe. He had heard about the Bowery and wanted to see it. So one night he set out on foot, but hardly had he gone a block down this notorious street before he was held up by thugs. Fortunately the assailants became frightened and fled before they could steal his wallet. But Holt proceeded on his way only to be set upon by other robbers who made off with his wallet. However he continued with only a few coins in his pocket when for the third time he was held up by gangsters who cleaned him out completely. But it was too late, he had accomplished his purpose—he had seen the Bowery.(10)

On another occasion, the factory hands organized a street parade and coaxed "Uncle Ben" into agreeing to lay aside his work for a few hours and lead the march. When it came time for the procession to start, however, Benjamin Holt had disappeared. A last minute search located him in his shop where he was dressed in his best clothes which were now covered with grease from his machines. He had forgotten all about the parade as he worked on the design of a new tractor.

When Benjamin Holt died on December 5, 1920, his achievements were widely recognized. The "Stockton Daily Record" observed that Holt's inventiveness had revolutionized farming and earth-moving practices in most countries of the world. His personal friends knew him as a friend of the common man, one who used his genius to provide machines which would lighten the burdens for both man and beast. The "Granite Monthly" in Concord, New Hampshire added that Holt had put farming on a higher plane of efficiency and contributed to the success of the Allies in World War I.(11) "Every California farmer knew Benjamin Holt," reflected the "Pacific Rural Press", and "for many years the Holt tractors outnumbered all others on the farms of California."(12) Long-time co-worker Paul E. Weston remembered that "Uncle Ben" was in a class by himself . . . He got the idea for the Caterpillar and sprang it on us. I thought it was crazy."(13)

Today a modest plaque in front of Holt's former residence in Stockton at 548 East Park Street identifies the homesite and concludes:

> The Caterpillar Tractor, born of the brain of Benjamin Holt, straightened roads, leveled valleys, flattened mountains, stored waters, cleared jungles, and served our National Defense. Its crawler track has carried the burdens of mankind and produced food for the mouths of the world.

This is the story of the man, his machines and his dreams. It is also a history of a company that began as a farm implement manufacturer in a small California town and grew over a course of a century to become on of the nation's largest exporters of heavy machinery.

Chapter 2

THE HOLTS MOVE WEST

Benjamin Holt's first American ancestor, Nicholas Holt, left England with his family in 1635 and settled near Salem, Massachusetts.(1) After residing in this bleak, rocky place for nine years, Holt moved inland where he became a prosperous farmer near the village of Andover, Massachusetts. Nearby pine forests provided lumber for daily essentials; including fences, furniture, fuel, canoes, wagons, fortifications, dishes, baby cradles, and tool handles.

Nicholas' descendants moved north to the forests of New Hampshire where Benjamin Holt's father, William Knox Holt, was born in 1811 in Loudon, a village eight miles northeast of Concord in the Merrimack River Valley. William and his second wife, Harriet Parker Ames, had eight children. Benjamin Leroy, the seventh, was born on January 1, 1849. All the Holt sons learned sawmill and woodworking operations in their father's hardwood lumber mill.

After completing elementary school, Benjamin attended Tilton Academy near Concord and a Baptist school at New London, New Hampshire. His older brother Charles, with whom he was later a business partner, attended the New London Literary and Scientific Institution (today Colby-Sawyer College).(2) Upon completing school, the Holt brothers worked in their father's business in Loudon from 1865 to 1873, when they moved the operation to Concord, a transportation center close to hardwood forests with plenty of skilled wood-working craftsmen.

Concord was the home of the famous Concord stage coach, a vehicle familiar today from "Westerns" and Wells Fargo Bank advertisements. First made in 1826 by Lewis Downing and J. Stephen Abbot, these coaches gained a wide reputation as being carefully built and smooth riding. The body was cradled in leather bands which cushioned the shock of traveling on rough roads.

Although the Abbot-Downing Company was Concord's major industry in 1890, the factory was able to produce only 50 stage coaches annually with its 250 employees. Priced from $775 to $1250 each, these coaches were sold primarily to stage line companies and resort hotels. (3)

Most New Englanders still needed inexpensive lumber wagons, buggies, and carts for transport, and it was here that the Holt brothers found their first big market. In the 1850s they sold lumber, but added wagon wheels in 1865 and later expanded the business to include the manufacture of wagons, buggies, sleighs, and a variety of hardware items.

After moving to Concord, the family bought the Benjamin F. Caldwell carriage works, a choice property located on the main road running to the large town on Manchester and contiguous with the railroad yards of the Boston and Maine Railroad. By 1890, the Holt factory had expanded to include a three-story brick structure, a large warehouse, a wood-working shop, and a powerhouse with a 100-horsepower steam engine. An elderly Concord resident recalled that "farmers always took their wagons to the Holt brothers each spring for fixing. Never went anywhere else." (4)

The successful Holt Bros. Company carried on business in Concord from 1873 to 1945, managed first by its founder, William Knox Holt, then by his sons, Benjamin, A. Frank, and Charles, and finally by a grandson, Edgar Holt. Beginning in 1864, however, four of the Holt sons left this productive wagon and carriage business and moved across the continent to California.

In the spring of 1864, Benjamin's older brother Charles Henry went to New York City where a sea captain offered him a job in San Francisco. When Holt arrived in the Bay city, however, he discoved that his would-be employer had gone bankrupt. Undaunted, he secured a teaching position in Northern California's Humboldt County. With the extra income earned by keeping books for a general store in Hydesville in the evenings for two years, he saved $700. Returning to San

The site of the Holt wagon factory in Concord, New Hampshire where the Holt family carried on business from 1876 to 1945.

William K. Holt and his three sons, Charles, Frank and Benjamin purchased the Benjamin F. Cadwell carriage works in Concord in 1876 for $20,000. The initial purchase included a wooden blacksmith shop, a brick machine shop, a woodworking building and a horse shed. In 1880 the factory had been expanded to include a three-story brick structure, a sawmill complex and a powerhouse which contained a 100-horsepower steam engine. Today these buildings are used for an automobile repair shop.

Francisco in 1865, he established C.H. Holt and Company in a two-story brick building on Beale Street. (5)

With the aid of brothers William Harrison and A. Frank, who joined him in 1871, Charles imported and sold lumber for building boats, mining machinery, wagons, and stage coaches. In 1880, the company issued a 106-page catalogue which included a wide range of items such as buggies, light spring wagons, Concord white-oak wheels, and hardware and carriage supplies. While claiming to carry the largest stock in this line on the Pacific Coast, the management assured, "All wheels are made of thoroughly seasoned stock and warranted to stand any climate. When ordering, call for Concord wheels, and accept no others, for they are the best."(6)

The Holt San Francisco business proved successful, and according to William's son Pliny E. Holt, Charles began it by importing hardwood from his father's New Hampshire sawmill. Paying $30 per thousand feet and an additional $30 for the freight to San Francisco, he was able to sell the lumber for $180 per thousand feet, which he insisted his customers pay him in gold, and made a 300 percent profit. Then he sent the gold to New England where $180 in gold brought $360 in currency, thus bringing a total profit of $300 on a shipment of wood which had cost him $60. (7) (This practice of buying supplies for paper money and selling them for gold was common prior to 1879 when the Federal Government made paper money equal in value to either gold or silver dollars.)

Almost 20 years after Charles began his career in the West, younger brother Benjamin left Concord, traveled across the country by the newly completed transcontinental railroad, met with his elder brothers, and determined with Charles to start a wheel company in Stockton, some 80 miles east of San Francisco. Brother A. Frank replaced Benjamin in the New Hampshire business.

The reason for the emigration of the Holt brothers to California may have been the dwindling of the hardwood timber supply in New Hampshire. Perhaps stiff competition with the Abbot-Downing Company and the Concord Axle Company was keeping profits marginal. Or perhaps the economic opportunities for utilizing the vast natural resources west of the Mississippi enticed those willing to take risks.

Undoubtedly, the Holts were also exposed to the advertising of businessmen who portrayed the West as a utopia in order to enhance their own profits. Railroad officials published pamphlets extolling the advantages of life in California where fertile soil would produce 60 bushels of wheat to the acre and grow watermelons three feet long. One New England newspaper editor in 1873 urged the readers to leave care behind and go to California, "a modern Garden of Eden, splashed by a silver sea, sprayed with golden color, ripe with roses, and purpled by the shadows of tropic palms." (8) Another enthusiast declared that California children grew up to be more beautiful than any race on the face of the earth and that even the mosquitoes were harmless and bed bugs unknown. As for the real or imaginary reasons Benjamin and his brothers moved west, perhaps, as on Holt family member recently suggested, "There were just too damn many Holts to feed."(9)

Chapter 3

WHEELS FOR THE NEW WESTERN AGRICULTURE

When Benjamin Holt and his brother Charles established a business in Stockton in 1883, their decision to do so was unequivocal. The choice of Stockton as the best place to set up a new West Coast operation seems to have been carefully considered, probably after oldest brother William had investigated the place firsthand. Several factors must have influenced their decision.

In 1849, the Gold Rush had swelled the village—named for Commodore Robert Stockton who commanded the United States naval forces during the Mexican War—into a busy canvas town serving a large transient population. Boasting good river connections to San Francisco on the west and wagon routes to the mines on the east, Stockton was described by historian Hubert H. Bancroft as a lively port where gambling, drinking and dissolute behavior were carried to excess. (1)

In just a few years, however, Stockton had become more than a jumping-off point for the gold fields and a place for miners to await cessation of the winter rains. Situated in a good location near the mouth of the San Joaquin River, Stockton had access to the river traffic of the entire Central Valley, the 475-mile long trough cradled between the Sierra Nevada and the Coast Range.

Five combines pulled by a total of 165 mules in Walla Walla, Washington in 1904. All are side-hill combined harvesters patented by Holt Manufacturing Company in 1891. These machines made it possible to harvest wheat in fertile hill country that was previously uncultivated.

Two-thirds of California's tillable land lies in this valley, a fact which became critically important for the fledging town Benjamin and Charles Holt selected to be the site for their new factory.

Prior to the 1850s, agricultural production in all of California had been exceedingly limited. For many centuries, native Americans had maintained themselves on a diet of acorns, wild berries, fish, and wild game. In the eighteenth century, the Spanish and Mexicans introduced livestock and a wide variety of grains, fruits, and vegetables, and the Russians at Fort Ross experimentally raised wheat and barley until the late 1830s. For the most part, however, the subsistence farming in the area was barely able to provide enough food for the Anglo settlers, who in 1848 did not exceed 13,000 people.

The discovery of gold brought 366,000 people into the state within a 12-year period, resulting in a severe food shortage and extreme inflation. In 1850, apples sold for $5, eggs were $50 a dozen, and whisky rose to $30 a quart. On November 30, 1852, the Stockton "Journal" noted anxiously that wheat was selling for $7.50 a bushel and flour was $1 a pound. "When men are starving they go to extremes," the paper warned. "Playing with bread before a hungry community whose wages are consumed by it will be like putting one's head into a lion's mouth. He may not crush it—then again he may."(2)

These food shortages were accentuated by the belief among the first immigrants that crops could not be grown in California's Central Valley. In the long summer months, the dry vegetation, the cracked earth, the absence of rains, and the torrid temperatures discouraged them from attempting to raise their own food. While early explorers such as Father Pedro Font and John C. Fremont praised the land's potential for pasturage and crops, a contributor to the "Overland Monthly" in November 1866 bemoaned the pervasive brown color of the valley in summertime, and its lack of woods and streams to break the monotony of blinding dust which buried wagon wheels to the axles. (3)

Although nature seemed to work in reverse, turning grass green in winter and brown in summer, farmers soon discovered how to raise grain by plowing and seeding in November after the beginning of the fall rains. As a a result, the food shortages of the early 1850s were fully overcome by the late 1860s, and grain buyers were bringing Central Valley wheat and flour downriver from Sacramento and Stockton to San Francisco for export abroad. Wheat raised in California increased from 17,000 bushels in 1850 to 44,000,000 bushels in 1890, putting California second in the United States in wheat production. (4)

Primary among the individuals who sparked the growth of Stockton as an industrial and agricultural center was Austin Sperry, a dry goods merchant from Vermont. After earning $500 from working 23 days in the placer mines near Stockton, Sperry shrewdly opened a grocery store in town. When the building burned in 1851, he built a mill to grind barley for the horses and mules that hauled freight to the mines. From this early

The Holt factory site in Stockton in 1887. By 1887, the Stockton Wheel Company had been expanded to include 5 buildings: a main office building, a warehouse, a paint shop, a blacksmith shop and a car and machine shop. By 1900, the Holt Manufacturing Company plant covered eight blocks in Stockton.

Plowing with mules in California. Photo taken near Vernalis, California, 1903.

The Stockton gang plow built by the H.C. Shaw Plow Works of Stockton featured small shares with little curvature of the moldboards because the ground in the Central Valley was easy to work. Usually four horses could pull a four-bottom gang plow. Each plow share was sharpened on both the top and the bottom, thus, when one edge got dull the share could be turned over or reversed to use the other side.

mill grew Sperry Flour Mills, which merged with six other milling firms in 1892 to dominate the flour milling industry on the Pacific coast. Moving its headquarters to San Francisco, Sperry and Company continued to expand, buying the Starr Mills in Vallejo in 1895 and the Port Costa Mill in 1910. During World War I, Sperry shipped flour in 10,000-barrel lots to the Allies in Europe. A huge sign on top of the San Francisco Ferry Building carried the Sperry Flour message, "Drifted Snow," and when General Mills Corporation bought out Sperry in 1928, they continued to feature the well-known Sperry label. (5)

The profitable San Joaquin Valley wheat trade, which began in the 1870s, depended upon advances in agricultural technology that made large-scale farming feasible. As early as 1872, the "Pacific Rural Express" described a 36,000-acre wheat ranch in San Joaquin County with fields 17 miles long. Men driving 40 horses and ten gang-plows moved down the furrows, stopping midway for lunch and camping overnight at the far end of the field before returning the next day. In 1880, Dr. Hugh Glenn's farm in Colusa County's Sacramento River Valley included 66,000 acres which produced a million bushels of wheat filling 12 ships dispatched to England. Forty men on horseback patrolled Glenn's fields, firing shotguns to keep wild geese from destroying the crop. (6)

Close to the bonanza wheatfields, Stockton had become an important city for the manufacture of farm implements by the time Benjamin and Charles Holt decided to locate their own manufacturing business there. As early as 1849, Monroe Robertson made wagons and repaired equipment at his blacksmith shop. In the same year Paige and Webster began manufacturing

Stacking grain using a Derrick Fork to lift the harvested grain from a barge and drop it on top of the stack. Artist's sketch taken from Thompson and West, "History of San Joaquin County, California," 1879.

plows at Main and El Dorado streets. One of the clerks in this company H. C. Shaw, bought the business in 1852 and continued producing plows. By 1886, 20,000 of his plows were in use, making the name "Stockton Gang Plow" known throughout the United States. (7) This Stockton firm, one of the oldest in the state, manufactured farm implements under the Shaw name until 1940.

Some Stockton-made plows were designed and manufactured specifically for use in the Central Valley. Because the arid summer months killed the grass crop, the tough, thick, virgin sod characteristic of the Midwest never formed. As a result, heavy breaking plows were unnecessary for cultivating new Valley land, and lighter plows were able to be pulled through soil devoid of roots, stones and gravel. Whereas Midwestern farmers usually used five horses to pull a two-bottom plow, Central Valley farmers needed only four horses to draw a plow with four plowshares.

Planting was also easier in the Valley. Wheat farmers in other states traditionally used a seed drill, which pressed the seed into the ground, but Valley farmers broadcast the grain from a seeder bolted directly to the plow. As the seed was sprinkled on the ground, it was covered by the plow and harrow in one quick operation. These special Valley planting implements were built by the Matteson and Williamson Company of Stockton, Marcus C. Hawley and Company of San Francisco, and Baker and Hamilton Company of Sacramento. By the early 1880s, Stockton led all other California cities in the manufacture of farm machinery. (8)

This fact alone may have been responsible for the Holt brothers' move to the thriving agricultural and industrial community in 1883. Certainly both Charles and Benjamin were well prepared to engage in their proposed new business. Aged 40 and 33 respectively, they already possessed between them—30 years of experience as wheelwrights and wagon manufacturers. In addition, they had enough capital to establish the Stockton Wheel Company without going heavily into debt. Shortly after their arrival in the city, Charles and Benjamin; apparently as previously agreed upon, bought out the financial interests of their two olders brothers, A. Frank and William Harrison. Charles assumed responsibility for the business end of the enterprise, and Benjamin took on the invention and production side of the work.

On land at the corner of Church and Aurora Streets, the Holt brothers built a three-story brick factory building. An additional frame building served as a boiler and drying shop where the wood used in making wagon wheels could season for three or four years. (9) The "Stockton City Directory" of 1884 carried one of the firms' first advertisements, announcing that the Stockton Wheel Company manufactured Sarven's Patent wood hub wheels, bodies, and gears in a factory near the Central Pacific Railroad Depot.

Soon Charles and Benjamin expanded their line of products to include blacksmith supplies such as trip hammers, hardware, harnesses, mining ore cars, picks, and shovels. Initially they employed 25 men in their shops, but as business rapidly increased, they added 42 traveling salesmen to cover the Central Valley and Northern California.

The Holt brothers' decision to locate their new buiness in Stockton was a shrewd move. Rapid growth in the Stockton area economy suggested continued prosperity. Laid out on land as level as a billiard table, Stockton claimed substantial homes, large trees, and a multitude of flower gardens. From a population of 10,289 in 1880, Stockton increased its numbers almost 50 percent by 1884.

The city site also claimed superb transportation facilities. It bordered the Stockton Channel, which ran three miles north to join the San Joaquin River leading to San Francisco 90 miles away. In 1883, more than 1,000 vessels departed from Stockton carrying freight and passengers between the two cities.

Likewise, railroad service out of Stockton was excellent. The Central Pacific line ran from Sacramento south to Stockton and then west to Tracy and the San Francisco Bay Area. A switching Y at Lathrop sent a branch line down the San Joaquin Valley through the tiny towns of Modesto, Merced, and Bakersfield. The Stockton and Copperopolis Railroad originated in Stockton, extending eastward 30 miles to Milton with a branch running from Peters to Oakdale.

In Stockton itself in 1883, streetcars connected the railroad depot with most of the principal streets, telephone service was available, and a new electric company provided lighting for businesses but not yet for residences.

Undoubtedly, the Holts were predisposed to locate in Stockton because Stockton ranked second only to San Francisco as the state's leading manufacturing center. The city had the largest number of farm machinery factories in the state. The Globe Iron Works, established in 1858, built steam engines, mining cars, steamboat machinery, and farm implements. Similiarily, the Stockton Iron Works, opened in 1868, produced castings essential to manufacturing farm machinery. The Matteson and Williamson Company, the Houser-Haines Company, and the Shippee Harvester Works already produced threshing machines, headers, and harvesters when the Holt brothers joined them in Stockton in 1883. (10)

The establishment of the Stockton Wheel Company by the Holt Brothers immediately added to the city's reputation as a center for the manufacture of wagons. William P. Miller had begun the tradition in 1852, building large freight wagons for hauling supplies to the mining camps. Twenty-eight feet long and eight feet high, with rear wheels seven feet in diameter, these "Stocktonian" vehicles weighed 5,000 pounds and sold for $1,000. Another Stockton resident, M.P. Henderson, built sturdy wagons for the Butterfield Stage Company. Carrying four tons and priced at $1,200 each, many of these heavy-duty wagons went to Death Valley where they were loaded with borax and pulled by 20-mule teams.

As the years passed, the demand for wagons in the Central Valley remained unfalteringly strong because of the amount of freight to be hauled out of the 25,000,000-acre growing area. With perfect drainage and almost two-thirds of the acreage dropping in elevation an imperceptible one inch per mile—making it perfect for mechanized planting and harvesting—wheat crops averaged 25 to 35 bushels per acre, double the national average.

During the California wheat boom, from 1860 to 1890, the grain poured out of the Central Valley like gold from the horn of plenty. In the north the harvest from Glenn, Colusa, Yolo, and Solano Counties moved on barges down the Sacramento River to the shipping terminals at Vallejo, Port Costa, and San Francisco. In the south, grain from Stanislaus, Tuolumne, Fresno, and San Joaquin counties descended the San Joaquin River and then went to the same terminals. The wheat from San Joaquin County alone in 1882 totaled 4,500,000 bushels. As early as 1869, 240 cargo ships loaded with Central Valley wheat departed San Francisco Bay carrying one-third of the nation's entire wheat exports. (11)

Over 95,000 sacks of grain stacked in sheds and on the ground at Ione, Oregon in October 1900. Each sack holds about 135 pounds of grain. Because of limited rain in the Central Valley of California during the summer months, grain sacks were often left in the fields or along fence lines for transport at a later time.

Hauling grain to the flour mill. Artist's drawing 1879.

The first Holt factory buildings in Stockton were located at the corner of Church and Aurora Streets. These were constructed in 1883. The three-story brick building measured 100 by 46 feet, and carried the advertising, "Stockton Wheel Co".

Warehouses located on inland waterways reportedly held two million tons of grain, and at times these facilities were inadequate. As a result, sacks of wheat were stacked in fields, along fence lines, and on warehouse platforms. Occasionally, wagons loaded with wheat lined up for a mile at terminals, sometimes waiting three days before being able to unload.

The enormous bulk of farm commodities hauled from Central Valley farms to market terminals was transported on wagon wheels. Materials for the building of new towns and for daily living in mountain and valley towns were also hauled by wagon.

Experienced in the art of making wagon wheels, the Holt brothers were highly successful in Stockton's wagon industry. Choosing woods that combined strength and light weight, they favored elm and black locust for hubs, oak and hickory for spokes, and white ash for "felloes," the wedge-shaped sections forming the outer circumference of the wheel.

Adapting quickly to the needs of people in the Central Valley, Charles and Benjamin augmented the two buildings at the Stockton factory with three more by 1887. The plant then consisted of a main office building, warehouse, paintshop, blacksmith shop, and machine shop. Their wagon wheel business continued on as late as 1910 (after the arrival of automobiles and trucks) and the company sold $35,000 worth of wheels, spokes, rims, felloes, and hubs. (12)

Ambitious and creative, Benjamin and Charles had more than just wagon wheels on their minds. Soon after arriving in Stockton, they turned their attention to the agricultural potential of their new home and their efforts ultimately changed the way in which farmers in California and across the United States harvested the bountiful fruit of the land.

Plows with grain seeder-box attachments were common in California agriculture because the plowing and seeding could be done in one operation. The seed grain fell on the ground where it was covered by the turned earth or by a harrow pulled behind the plow. Grain could also be broadcast from a wagon seeder and harrowed in the soil. This five-bottomed plow was manufactured by the Matteson and Williamson Company of Stockton, California. It has been restored and is now on exhibit at the Pioneer Museum and Haggin Galleries in Stockton.

Chapter 4

THE ARRIVAL OF THE COMBINED HARVESTER

On a carriage ride near Stockton on a summer day in 1884, Benjamin Holt and his companion passed farmers cutting wheat with a "header" machine and stacking the grain for threshing later in the season. A man of few words, Holt remarked, "That method requires too much work. Harvesting and threshing should be done in one operation."(1) Returning home, he set out to design and construct a practical machine that would complete this two-fold task.

Holt was not the first to point out the age-old inefficiency of harvesting grain, and to experiment with building a combined harvester machine that would first cut the tops off wheat stalks and then beat the heads to remove the grain kernels. In fact, many prototype machines had already been designed by individual mechanics and tried out in California by 1883. Because Benjamin and his brother owned factory facilities, however, they were in a position to develop the manufacturing of combined harvesters and write a new chapter in the story of agricultural advancements.

Technological progress in farming had moved exceedingly slow. Colonial Americans cut grain with a scythe, gathered sheaves by hand, and threshed with a flail almost exactly as farmers did in Biblical times. In the late 1780s, Andrew Mielke of Scotland invented a threshing machine that used a drum with spiked teeth revolving inside a set of cylinders with stationary teeth to knock the kernels out of the grain heads. This became a basic feature of virtually all subsequent threshing machines. (2)

The first threshing machines in the United States required the threshing cylinder to be turned by hand. By 1820, these "ground-hog" threshers were turned by horse-powered treadmills. As newly designed plows and reapers enabled farmers to increase grain production, Hiram and John Pitts of Winthrop, Maine, invented an improved machine in 1834 with an attached fanning mill which separated the grain from the straw and chaff. (3)

The flail wielded by manual labor provided the first power for threshing grain in colonial days. Original drawing provided by the J.I. Case Company of Racine, Wisconsin.

"Groundhog" threshing cylinders were in use by 1825 in the United States. These spiked cylinders knocked the kernels out of the grain heads. The cylinders were first turned by hand; then by a team of horses in a tread mill.

From 1830-1870, horse power machines driven by 4 or 5 teams provided power for threshing in most grain growing states in the United States.

These early threshing outfits were driven by sweep-power machines pulled by 8 to 14 horses walking in a circle. The machines incorporated the principle of a ship's capstan. With long spokes radiating from a central hub bolted to a cog wheel, the power was carried through a set of tumbling rods which provided rotary motion to the cylinder of the threshing machine.

Difficult to put in motion, these sweep powers depended on horses specially trained to exert slow, steady pull, and because the teams walked in circles around the hub, they were prone to develop sores from being pulled to one side. Some concerned citizens suggested the Society for the Prevention of Cruelty to Animals should intervene. (4)

While threshing work was hard on horses, the immense amount of hand labor required to harvest grain was also grueling. In the Midwest, most grain was cut by grain binders, which left bundles scattered on the ground. These bundles then had to picked up by hand and clustered in shocks for protection against inclement weather and for drying before threshing. This stoop labor was performed during the hottest months of the year. Alternately, if the grain were cut with header machines, it then needed to be pitch-forked into large stacks to await threshing— another laborious task.

Farm families usually hired migrant workers to complete this work. After shocking the grain, these men frequently found jobs with the owners of neighborhood threshing outfits. During shock-threshing, each bundle-hauler drove his team and rack out into the field, pitched on a load of grain, hauled the wagon to a stationary threshing machine, and pitched the bundles into a feeder. The loading and unloading of these racks was carried out in perfect rhythm, and each bundle-hauler had to keep up with this routine to prevent the threshing machine from running idle. Since as much grain as possible was threshed during favorable weather, dawn to dusk hours prevailed. Threshermen often quipped that they worked a nine-hour day—nine in the forenoon and nine in the afternoon. When one thresherman was asked if he was going to stop work to attend his wife's funeral, he reportedly replied, "No. A man can get a wife most anytime, but threshing comes just once a year." (5)

Threshing crews worked under trying conditions. Their shirts were permanently black with sweat and dirt, while winds parched lips and filled eyes with dust. Mosquitoes, gnats, and flying ants harassed the crews relentlessly.

The "Houser" and the "Prince" advertised by the Stockton Combined Harvester and Agricultural Works and the Matteson and Williamson Manufacturing Company in the "Pacific Rural Press" on April 2, 1892.

Farming on the huge wheat ranches in California accentuated the demand for machines, animals, and manual labor. Dr. Hugh Glenn, a Virginian who began buying land along the Sacramento River in Colusa County with $10,000 won in a poker game, raised a million bushels of wheat in 1880 on his 66,000 acre ranch. This necessitated an investment of $125,000 in farm machinery, $185,000 in horses and mules, and $100,000 in buildings, including 32 houses, 27 barns, and 14 blacksmith shops. Glenn's machinery line-up included six steam engines, six threshing machines, 60 headers, and 180 header boxes or barges. One reporter described the Glenn Ranch in 1880 as a vast, unbroken sea of yellow grain—"huge, calm, and exceedingly beautiful." The writer duly noted that Glenn bought 1,000 horses and mules and hired 900 field workers and a host of blacksmiths to put the machinery in order. Overseers rode about on horseback giving instructions to workers, much as if preparing a general's army for battle. (6)

Threshing began as headers entered a field, cutting a circle of about four acres. Then the threshing machines and steam engine set up in this circle, while the headers completed harvesting a square mile of grain. Each threshing team consisted of one threshing machine with 25 header wagons, 70 to 80 men, and the same number of horses and mules. Most men were dressed in brown canvas jumpers and overalls, with broad-brimmed straw or felt hats to protect them from the sun.

In the fields, working conditions were miserable as the sun burned into the ground until clouds of grey dust rose at the lightest step. One member of a threshing crew on the Glenn Ranch in 1877 wrote:

> At present the mercury stands at a hundred degrees in the shade, and I long for a Nebraska thunderstorm to break the monotonous weather. Mr. Glenn is considered the greatest wheat grower in the world, but I will admit that this is not home sweet home. Wages by the year are $360. Wages in harvest time are $1.50 to $2.50 per day. It is so hot I can't get my thoughts so I remain without a struggle . . .(7)

Sometimes workers slept in sheds or straw stacks, but usually they just spread their blankets on the ground under the stars.

Feeding the men was simplified in 1876, when J.B. Greene of San Joaquin County conceived the idea of a cookhouse on wheels which could be moved about with a team of horses. Bringing meals to the men rather than men to the meals saved valuable work time. One observer recalled a mealtime scene:

> At noon a huge van is driven upon the field laden down with meat, vegetables and pies, all well cooked and very palatable. Farmhands, like fishermen, nowadays are epicures. This van is so constructed that the sides form broad tables. The cooks who serve stand in the body of the wagon, and the diners range themselves around the outside. All are sheltered by a screen or canvas overhead. (8)

Chinese cooks prepared meals with a minimum of spoilage using home-made evaporative coolers. Meat was

A Holt combined harvester operating in the field. Note the driver perched on the elevated platform. The man by the tiller raised and lowered the cutting platform according to the height of the grain.

delivered by butcher wagons making 25-mile rounds, and dishwashing was claimed to be as sanitary as in a city restaurant. (9)

Although time-saving, these cook-wagons occasionally introduced new problems. The owner of a farm adjacent to the Glenn Ranch recalled that sanitation was almost non-existent and that only frequent moves of the camp sites prevented serious epidemics.

The threshing hands included local farm boys, Civil War veterans, ex-miners, cowboys, dock workers from the warehouses of Port Costa, and immigrants from Ireland, Germany, and the Scandanavian countries. Many were crack shots, skilled horsemen, and as rugged as the country in which they found themselves.

On Saturday nights header wagons from the Glenn Ranch dropped the hired hands in the streets of the town of Willows where they proceeded to relax with uninhibited gusto. One side of Main Street boasted 25 saloons and the other the houses of 75 prostitutes. By Sunday night, when the wagons returned, the alleys held many of the celebrants. Poker was a favorite indoor spot, and an occasional card sharp made off with the threshing wages. At times guns lay on the tables beside the chips. (10)

For all the success of businessman-farmers like Dr. Glenn, forces were at work which reduced the huge size of the farm labor force and ultimately eliminated most back-breaking labor from the harvest fields. One innovation in threshing that came into vogue in the 1870s in the far West involved the use of a derrick and net rather than laborers with pitchforks to unload barge-wagons. Also, when threshing from a stack, the derrick was used with lift forks to carry wheat to the threshing machine. This also represented a significant reduction in work over pitching grain by hand.

It was the advent of the mechanical combined harvester, however, which eventually banished the curse of the pitchfork from grain farming. The combined harvester also reduced the cost of harvesting from $3.50 per acre in 1875 to about $1.50 an acre fifty years later, primarily by eliminating human labor.

The idea of combining the harvesting and threshing of grain into one operation was first incorporated in a patent by Samuel Lane of Maine in 1838, but apparently no machine was built. Mrs. John Hascall of Flower Field, Michigan, also deserves some credit for the invention of the combined harvester. One morning in 1834 she told her husband that she dreamed about a large machine drawn by horses which went over a wheat field, cutting and threshing the straw in one operation. That afternoon a neighbor with an inventive mind, Hiram Moore, drew a rough sketch of such a machine and later built a model, secured a patent, and used it in the field trials the following year. Although Hiram Moore built five different machines prior to 1853, their success was severely limited because Midwestern conditions required cutting the grain before it was dry enough to be threshed properly. (11)

In 1853, one of these combined harvesters was shipped around the Horn to San Francisco and taken to John M. Horner's farm near Mission San Jose in Alameda County. In the summer of 1854, it successfully harvested 600 acres of wheat. Two years later, an operator failed to oil a bearing, which caused it to get so hot that it started a fire and destroyed the machine. (12)

Undaunted, Horner built another combined harvester in 1859 which cut, threshed, and sacked 16 acres of wheat per day. Three more were constructed and used in Santa Clara and San Joaquin Counties in 1868. Fears that the new machines would put men out of work prompted arsonists to burn one of them in 1869, an event which was followed by offers of a $500 reward for the arrest and conviction of the men responsible. (13)

Working along similiar lines in 1862, William Marvin and H.H. Thurston of Stockton attached a threshing machine to a header and harvested grain near the city. Yearly improvements were made, until in 1867 the "Stockton Independent" declared, "This machine is considered one of the greatest laborsaving inventions ever introduced." (14)

Despite earlier individual experiments around the United States, the first successful use of combined harvesters occurred in California where the arid summer

In 1898 Holt combined harvesters ranged in price from $1,650 to $1,700 with headers cutting swaths from 16 feet to 20 feet wide.

months devoid of hail or high winds, permitted the grain to get dead ripe in the field before harvesting. The existence of large-scale farming and the scarcity and relative costliness of farm laborers served as additional imperatives for developing labor-saving machinery.

Reflecting their origin in the stationary threshing machines long in use, these new harvesting mahines were called "traveling threshers" or "traveling thrashers". (Benjamin Holt's first patent of August 6, 1889, was registered under the title, "An Improvement on the Traveling Thrasher.") Some people called the machines "combined headers and threshers", which accurately described their dual function, but by the 1890s the name commonly used was "combined harvester." When these machines later came into more general use, the term was reduced to the single word, "combine." (15)

With a ready market for the new invention, manufacturers raced to produce them. David Young and J. C. Hoult built several combined harvesters in Stockton in 1876, calling them the "Centennial Harvester" because of the United States' centennial year. When the Matteson and Williamson Manufacturing Company produced several combined harvesters in the same year, the company may have been the first to manufacture more than one of the same model. The Houser-Haines Company built their first combines in the same city in 1880, while the Shippee Harvester Works followed in 1883. In 1884, the Shippee and Houser companies merged to form the Stockton Combined Harvester and Agricultural Works, promptly buying up the patent rights of several other local builders of harvesters. (16)

The early combined harvesters attracted considerable attention. Jacob Price, an engineer in San Leandro, California, made a trip through the San Joaquin Valley in 1885 to see them at work. He noted that the Houser, Shippee, and Young machines were used most extensively, and remarked that at work, they reminded him of ships at sea—ponderous juggernauts sweeping away all before them and dropping off sacks of grain like dead men cast aside. While marveling that such monsters worked as well as they did, Price was convinced that they would eventually supersede all other methods of harvesting grain. "The Stockton mechanics are to be congratulated for their enterprise and intelligence," he concluded, "and I believe Stockton to be the most favorable place in the United States for the development of new agricultural machinery." (17)

Keenly aware of the propitiousness of the times, Benjamin and Charles Holt quickly joined the cavalcade of combine manufacturers. Having been in Stockton only two years with their Stockton Wheel Company, they decisively bought several patents, expanded their factory facilities, conducted experimental work, and put their first combined harvester on the market in 1886. Boasting a 14-foot header and a 21-inch threshing cylinder and pulled by 18 horses, it was sold to Joe Bunch (later a Holt Company mechanic) who harvested 25 acres a day for 45 days. Bunch bought another Holt combine in 1887, which he operated for the next 37 years to harvest over 50,000 acres. In 1924, it was reported to be in good working condition. (18)

Observing these early combines at work, Benjamin Holt knew that improvements could be made in their design. He noticed, for instance, that they carried the power from the drive wheels to the moving parts of the harvester by means of a set of tight gears. These gears wore rapidly from the dust in the field and created extensive friction which required more horses to pull the machines.

Not only were these gears short-lived, necessitating costly repairs, but they growled so loudly that they frequently frightened the horses. On July 9, 1887, the "Pacific Rural Press" reported that 22 horses pulling a combined harvester near Visalia, California, had gotten out of control and thrown the driver to the ground. Two of the horses were killed and six more injured. A long-time combine operator recalled that a combine pulled by 33 horses might go along smoothly until a gear broke down:

> "As long as the animals heard the regular music of the harvester, all was O.K. When the tune changed suddenly, they would go away from there in a hurry, and the combine had to be rebuilt. Now and then a horse or two had to be shot, and a funeral or two of drivers during the season in the San Joaquin Valley was not unexpected. I recall 13 runaways in one season." (19)

To forestall the problem of runaway combine units, some early builders such as John Horner and L.U. Shippee placed the hitch for horses behind the machine so that they pushed it. When frightened, the horses tended to stop and back up rather than bolt ahead in dangerous fashion. Twenty-four horses in two rows of 12 each behind the machine created such a phalanx of animals, however, that the driver could barely control and guide them.

Having observed the negative features of the early combines, Benjamin Holt designed his first machines without the set of tight gears. This eliminated excessive friction as well as the growling noise it created. Holt substituted chain drives running from a sprocket bolted on the main drive wheel to a countershaft which in turn carried the rotary power to the cylinder, cutting bar, reel, drapers, and other moving parts of the threshing machine. These chains, called "link belts", were made by coupling individual links. If one link broke, it could be replaced quickly with another one, causing little expense or loss of time. The link belt resulted in a smooth-running, quiet machine which an 1889 Holt brochure advertised as:

> Indisputedly the simplest, lightest draft, most easily handled of any harvester ever introduced. Not a single runaway has occurred with our harvester with any serious result. When a team starts quickly, the heavy

clumsy gears make a terrific noise, frightening the animals beyond control, causing a general smash up. A sudden start on our harvester produces no such noise, not so much as an ordinary header. Should the team run, the chain parts and falls harmlessly to the ground. The chain can be adjusted and a new link substituted in five minutes.

A second engineering innovation in Ben Holt's combines consisted of a V-Belt that drove the threshing cylinder. Two V-pulleys, one on the countershaft and another on the cylinder shaft, carried a V-Belt made of leather. According to Holt employee Tom Luke, "The V-Belt on the early Holt combines was a three-ply riveted leather belt two inches wide and tapered to fit the V-shaped pulleys. To this belt were riveted . . . lugs built up of layers of sole leather gradually tapering to a point, giving immense contact with the V-pulleys." (20)

The engineering excellence built into the early Holt combines became apparent when in 1887, owners of Holt, Houser, Shippee, Meyers, and Young combined harvesters gathered on a wheat ranch near Lodi, California, for a field competition. Combines that clogged or "slugged" as they moved through exceptionally heavy grain were disqualified, leaving only the Holt and Houser machines to complete the test. Another contest near Stockton tested how much grain combines could cut and thresh in a fixed time period. Of five combines, the Houser took first place, and the Holt came in second.

A time capsule found in an iron column in a Holt factory building constructed in 1899 reveals that Holt first produced 13 combines in 1886, 134 machines in 1892, 61 in 1896, and 128 in 1899 for a total of 1,072 combines before 1900. By 1900, the sales of combines by the Holt Manufacturing Company (including the companies Holt purchased in the 1890s) were greater than those of all competing firms, and by 1916, some 6,000 Holt combines harvested an estimated 90 percent of all the grain on the Pacific Coast. From 1886 to 1929, Holt sold 14,111 combined harvesters, representing almost complete domination of the market. Benjamin Holt had made Stockton the combine capital of the world. (21)

Benjamin Holt's post-1900 machines were more complex in design than the early models. Moving down the field, the combine's sickle bar and reel toppled the standing grain onto a moving draper, or belt, which carried the grain up into a hopper. Here another draper moved it into the swirling cylinder of the treshing machines, where the kernels of grain were knocked out. Then a series of shakers, sieves, and fans separated the grain from the straw and chaff, after which a cleaner removed weed seeds before the grain descended into a hopper, ready for sacking. Carrying out these operations were gears, chains, belts, sprockets, augers, pulleys, and wheels churning away in a din of noise and a stifling cloud of straw, chaff, and dust. Lumbering across rough terrain or hillsides, the combines were an awesome sight.

When the Holt combines first went into use near Ritzville, Washington, in 1888, farmers traveled as far as 70 miles to see them perform, while a special train carried spectators from Pullman to Endicott, Washington, to witness the first use of a Holt combine in the famous wheat-growing Palouse country. The drivers of these combines, who cracked their long whips over six or seven rows of animals below, acquired status like that of the captains of riverboats or stage coaches. (22) A 1904 photograph of five Holt combines at work on the Drumheller farm near Walla Walla, Washington, shows 165 horses and mules hitched to the five outfits. One of the most dramatic views of farming ever made, this image was widely distributed on post cards, and in 1909 Benjamin Holt received one from a postal employee reading: "This postal is the largest seller ever gotten out in postal cards." (23)

Because combined harvesters drastically reduced the size of the labor forces required for harvesting and threshing, they had an unexpected emancipating effect for farm women, who were expected to cook 6:00 a.m. breakfast and 9:00 p.m. supper not only for their large families but for the shocking and threshing crews of 15 to 20 men. On some of the largest threshing outfits, a special cook car followed the operation, but this work usually fell to the farmer's wife, daughters, and neighbor women.

For weeks in advance of the harvest, women canned vegetables and fruit, dug bushels of potatoes, and spent days baking as many as 40 loaves of bread. Complaining that thresher meals were too elaborate, one farm woman suggested that only a few "simple" dishes be prepared for breakfast: coffee, cream, sugar, oatmeal porridge, bread, butter, jam, fried potatoes, ham and eggs, and fresh steak. (24) Another farmer's wife near Modesto, California, wrote to Benjamin Holt in 1908 to express her gratitude for the combined harvester, which had freed her from many long hours in a hot kitchen by reducing the number of men required for the harvest. (25)

Operating a combine now required a skilled crew of four or five men instead of 20 to 30 transients. George Reister, who farmed near Willows, California, ordered a $1,200 Houser-Haines combined harvester in 1904, which arrived on a railroad flat car and was unloaded by mule teams. His threshing set-up was typical for the era. In the field, a header tender, or a "header puncher" turned a tiller wheel to raise or lower the header platform. A separator man clambored along catwalks astride the combine, oiling the bearings every two hours, adjusting the sieves, and checking that grain did not ride over on the straw and tailings. A third man, called a "sack jigger," hooked the open end of grain sacks on a rectangular frame, then lifted a gate on the grain hopper to fill the sacks. Next, the seated sack sewer grabbed the two-bushel sack from the sack jigger, placed it over a

A Holt combined harvester on the Harris farm in San Joaquin County, California. Note canvas shade provided for the grain sackers.

cleat, grabbed the corner of the open sack to form an ear, put a double hitch around it with needle and twine, rolled the top of the sack over, and sewed it closed with nine to 11 stitches. He then placed the sack of grain on a small platform, or "dump board," which balanced on a fulcrum like a teeter-tooter. The first four sacks placed on the dump board remained balanced in position, but the fifth sack automatically tipped the platform, dumping the sacks in a pile on the ground. The dump board then flipped back to its original position. (26)

In the West, grain moved from field to market in sacks instead of in bulk because sacked grain could be left outside in summer without being damaged by rain. As a result, towering railroad elevators or "prairie skyscrapers" seldom marked the rural California landscape.

Most farmers liked sacking grain because good quality grain could be more easily separated from inferior grain. Many buyers favored sacks because grain left in storage bins tended to dry out and lose weight, bringing lower prices. (Buyers often left piles of sacked grain on waterfront docks to absorb moisture from the air, increasing the grain's weight and profitability.) Shippers also preferred to handle sacks, because bulk grain loaded in ocean-going vessels shifted position more easily in stormy seas which endangered the safety of the ship. Also higher insurance rates were charged for bulk grain shipments.

Sacked grain had drawbacks, however, especially high labor and handling costs, deterioration from sunburn, vulnerability to rodents, and leakage in handling. At harvest time, sacks were always in short supply, and in Hollister one year, several harvesting outfits had to shut down for lack of sacks. Manufactured for a time at San Quentin State Prison and sold by the state, sacks were increasingly criticized as unduly expensive. Grain sacks, woven from jute imported from India, usually cost from 7 cents to 14 cents each. In 1916, the "Pacific Rural Press" began publishing articles urging the elimination of sacks, and during World War I, the Holt Company began attaching grain bins to some of their combines, thus finally eliminating the need for sacks. (27)

Threshing was more than just expensive and hard work—it was dangerous. One veteran thresherman from the wheat lands of Washington's Palouse region recalled that wives anxiously bade their husbands goodbye when the men drove their horses to the rolling harvest fields. Combines clung perilously to hilly "farms on edge", and operators had to be ready to jump for their lives at any moment. Since many machines had no brakes, horses had to be kept moving; if they stopped going up a hill, the machine would roll backwards and pull everything with it to destruction. Even "little acts of God" like rattlesnakes, bumble bees' nests, or a sudden flight of a flock of game birds would start a team running. According to one observer, nerve and skill plus good and bad luck decided whether the driver would live to drive another day." (28)

To deal with the problem of hillside farms, Benjamin Holt designed a side-hill traveling harvester in 1891 (Patent No. 483, 449). Prior to this time, combines were usually confined to level ground, because on hillsides, grain would drift to one side of the machine. This caused the machine to clean the grain improperly or let it fall out on the ground. (The Holt Company claimed two to six bushels per acre were lost this way.) Benjamins Holt's ingenious side-hill combine made it possible to keep the threshing machine in a horizontal position regardless of the terrain. By designing the thresher so that each drive wheel turned within separate wooden frames, much like the wheel on a wheelbarrow, each drive wheel could be lowered or raised by adjusting the wooden frames. Operators made these adjustments by simply moving a small lever. Holt's new machines, which maintained a horizontal position on hills with inclines up to 30 degrees, could cut 35 acres a day. (29)

The first Holt side-hill combines were used near Oakdale in the San Joaquin Valley, but they soon reached the hilly regions of eastern Washington and Oregon. In 1896, the "Stockton Evening Mail" estimated that there were 125 side-hill harvesters in use.

Benjamin Holt took a keen interest in improving the combined harvester. Of the 47 patents granted to Holt between 1889 and 1921, 18 applied to improvements in the design of the combine. In addition, many patents granted Elmer E. Wickersham, Pliny E. Holt, William Turnbull, and E.F. Norelius, assignors to the Holt Manufacturing Company, pertained to combine inventions.

Although conflicting claims make it difficult to prove the origins of certain inventions, it appears that Benjamin Holt and his associates made a number of important mechanical innovations related to the

Holt 1898 catalogue designed to demonstrate the features of the "Side-Hill Combined Harvester." The first patent for the "Side-Hill Combines" was secured in 1891.

This "Side Hill Holt Combined Harvester" was manufactured about 1900.

combine. The Stockton Company claimed to be the first to:

- Replace gears with chain drives on combines (1886)
- Use V-belts to drive combine cylinders (1886-1887)
- Invent side-hill harvesters (1891)
- Use counter-balancing weights to facilitate the raising and lowering of the combine header platforms (1886)
- Use a single front wheel with a turntable for improved steering of combines (1886)
- Adopt the hinged header to facilitate adjustment of the sickle bar (1886)
- Build combines with a 50-foot sickle bar (1893)
- Design combines with power for the threshing machines derived from two drive wheels (1893)
- Build combines with steel separator shoes (1894)
- Adopt tapered shanks for cylinder teeth (1897)
- Attach auxiliary gasoline engines to combines (1904)
- Patent self-propelled combines (1908) and build self- propelled combines (1911)
- Manufacture combine attachments for picking up windrows or rows of cut grain (1926) (30)

This imaginative engineering helped make the Holt combined harvesters among the best in the world. Noted for their efficiency, reliability, and longevity, they cost from $1,550 to $1,750 depending on the size and model. Although they were not the lowest-priced on the market in 1900, they were backed up by one of the most reputable firms on the West Coast. The Holt Company's combine won a Grand Prize at the 1915 Panama Pacific International Exposition in San Francisco, and today an 1887 horse-drawn Holt combine stands in the Museum of Technology of the Smithsonian Institute in Washington, D.C. The machine represents not only what historian and Smithsonian Curator of Agriculture John T. Schlebecker termed the "final development of the heroic age of animal power," but a tribute to the people of the San Joaquin Valley who designed and manufactured labor-saving combines for use on the West Coast 40 years before they came into common use in the rest of the United States.

The versatile combine is now used for harvesting most farm crops including rice, soybeans, milo, and corn. It has displaced the sickle, sythe, reaper, binder, header, and threshing machine, as well as corn binders, corn pickers, and corn hullers. Noted agricultural engineer Roy Bainer unhesitatingly ranks the combined harvester as "the most important agricultural invention of the last 300 years," thus supplantin a position long held by Cyrus McCormick's reaper. (31)

This Holt combined harvester was pulled by 24 horses. Driver often drove the teams with only one set of reins.

Cook wagons were used to feed the threshing crews on the large wheat ranches in the Far West. These were pulled from place to place to accommodate the field hands.

Chapter 5

STEAM AND SMOKE ACROSS THE FURROWS

Not content with having produced popular combined harvesters a half-century before they found acceptance in the East, Western manufacturers like Benjamin Holt turned next to the task of designing a successful steam traction engine for use in plowing, logging, freighting, road-building and harvesting. Well over a decade before steam traction engines came into use in the Midwest and East, Western manufacturers like Benjamin Holt and his competitor Daniel Best designed and constructed the first practical self-propelled steam engines. This happened, not in the nation's Eastern industrial centers, but in Stockton and San Leandro, medium-sized California towns with less than 18,000 inhabitants.

The story of the steam traction engine begins in the mid-1880s when Benjamin and Charles Holt set up their manufacturing operation in the predominantly rural region of the San Joaquin Valley. Never satisfied with just one project on the drawing board or one kind of equipment being manufactured in their shops, they began building railroad cars for the Southern Pacific Railroad and supplying streetcars for several towns, including San Jose, Santa Cruz, and Stockton. When it became apparent by 1892 that railroad and streetcars

Portable farm steam engines first appeared in the United States in 1849. The engines were mounted on wheels so they could be pulled from place to place by horses. This was the first use of mechanical power for threshing. These portable steam engines were used in all states of the Union, not only for threshing, but for sawing wood, ginning cotton, grinding feed, and driving sugar mills. This is a picture of a J.I. Case engine built in 1869 and used in a threshing demonstration in Racine, Wisconsin in 1948.

were not highly profitable, they discontinued this line of manufacture. (1)

Unable to rely on nearby industrial suppliers, the Holts made their own castings and did their own machine work, blacksmithing, milling, and finishing. By the late 1880s, they expanded operations to manufacture boilers for steamboats and "Fresno scrapers" for building roads, levees, irrigation ditches, and railroad grades. The Holts also made mining cars, wagons, and drapers and sold lumber, leather belting, chains, iron, wrenches, forges, bench planes, mandrels, nails, bolts, harrow teeth, and oil and grease.

By 1900, the factory facilities in Stockton covered eight blocks. Among the formal factory departments were Benjamin Holt's experimental shop, a machine shop, foundry, boiler shop, carpenter shop, mill rooms, sheet iron building, pattern shop, drafting rooms, paint shop, wheel building, draper shop, and blacksmith shop. Switch yards connected the factory with two railroads passing through Stockton, while the Holt business office used Western Union and Pacific Postal long-distance wires. The "Stockton Evening Mail" on December 24, 1904, commented with obvious pride that, "The plant is one of the largest and finest equipped on the Coast, and every facility and modern device known to the industry has been installed regardless of expense." (2)

One important factor in the Holt's burgeoning industrial expansion was the stiff competition generated by other area businessmen. An unusually fierce rivalry developed between the Holt and Daniel Best families for control of the farm machinery business in the Far West, a battle which lasted 45 years and, like most prize fights of the day, went more than 15 rounds.

Born in Ohio in 1838, Daniel Best joined an Oregon Trail party in 1859 and prospected for gold in the Snake River country. Adventuresome and peripatetic, he worked as foreman of a sawmill in Portland, Oregon (until he lost the fingers on one hand), built a lumber mill at Tumwater, Washington, then moved in 1869 to Marysville, California, where his three brothers farmed. Observing that farmers had to haul their grain to town to get it cleaned at a cost of $3 a ton, he began in 1870 to build grain cleaners which could be moved from one farm to another. Best and his brothers received $200 a day for each of their three cleaners, which processed 60 tons of grain. (3) Moving to Oakland, California, in 1883, Best continued to manufacture grain cleaners until the machines began spilling over onto the streets near his shop. When local police objected, he purchased Jacob Price's Agricultural Works in nearby San Leandro for $15,000. Noting that farmers were beginning to attach grain cleaners to traveling combined harvesters this inventor built his first combine in 1885 and pursued a manufacturing career similiar to that of Benjamin Holt in Stockton. Between 1885 and 1925, his company manufactured 1,351 combined harvesters, approximately one-tenth the number produced by the Holt Manufacturing Company during the same period. (4)

Throughout the late nineteenth century, agricultural engineers knew that a better source of power than draft

Threshing wheat on the Dalrymple farm in the Red River Valley of Dakota Territory, 1878. Threshing in the Midwest was largely shock threshing. The grain was first cut with a binder and the bundles put into shocks. Later these shocks were hauled to a stationary threshing outfit. This drawing appeared in "Frank Leslie's Illustrated Newspaper" October 1878.

Threshing wheat in the Livermore Valley in California during 1898. The portable steam engines were pulled from one site to another by horses. Note the use of a wind stacker as an improvement over the straw carrier used in earlier years.

Best steam traction engine working for the White Pine Lumber Company located in the San Joaquin County of California. Photo taken about 1910.

animals was needed to move the large combines used in the Pacific Coast states. Draft animals were costly to maintain, big teams were unwieldy, and many horses died from sunstroke and exertion in the hot summers. Steam traction engines offered the only alternative to draft animals, and accordingly Daniel Best and Benjamin Holt set out to capture the new market for farm steam engines. This competition, beginning when Daniel Best built his first steam traction engine in the spring of 1889 and Benjamin Holt followed in 1890, took them through bitter legal battles, understandable today only when the importance of the technological leap from animal power to mechanical power is fully appreciated. (5)

It was the steam engine, in fact, that ushered in the Western world's industrial, commercial, and agricultural revolutions. The first practical steam engine appeared in England, when Thomas Newcomen installed one in a coal mine near Devonshire in 1712 for pumping water out of the mine shafts. Forty years later, James Watt and Thomas Boulton improved the Newcomen engine, which was limited to a vertical action, by building engines which attached the piston to a crankshaft, thus producing rotary power which could be belted to factory machines. Soon this type of rotary power was harnessed for railroads, steamboats, factories, and farms. (6)

In the United States, stationary steam engines were used on Southern plantations as early as 1807, 20 years before the operation of the first railroad locomotives in this country. By 1850, most large planters depended on these engines for belt work in threshing grain, sawing wood, ginning cotton, and crushing cane in sugar mills. (7)

Following the Civil War, farmers welcomed the manufacture of portable steam engines mounted on wheels which could be moved about by horses. Priced at about $1000, these engines developed 10 to 20 horsepower. They were used for custom threshing and other belt work in the grain-growing areas of the country. (8)

The Holt Manufacturing Company manufactured 132 wheeled steam traction engines from 1890 to 1911. A company catalog in 1898 illustrated their uses in freighting, logging, and harvesting grain. These engines were all equipped with chain drive systems.

A special breed of skilled mechanics operated these machines—the farm engineer. Responsible for operating the high-pressure boilers, making repairs, hiring threshing crews, looking after the crew's safety, and orchestrating one of the largest group activities in farm life, farm engineers were idolized by children who dreamed someday of riding with one hand on the throttle, a sharp eye on the engine, and an ear to the rhythmic "tuck-a-tuck" of the engine.

As America expanded westward, the demand for power accelerated as the scale of railroading, ranching, mining, lumbering, and farming operations increased. In 1884, on a bonanza farm in the Red River Valley of Dakota Territory, Oliver Dalrymple used 30 portable steam engines to thresh 30,000 acres of wheat. A contemporary observer near Casselton counted 53 steam engines in nearby fields, each threshing 1200 bushels a day. (9)

Convinced that threshing rigs made in the East were too small for huge bonanza wheat farms in California, agriculturalist Dr. Hugh Glenn asked his blacksmith, George H. Hoag, to build a threshing machine with greater capacity. The result was a 35-foot-long machine which, when belted to an Enright engine, merited this report in a California newspaper, "Willows Journal", on July 26, 1879:

> At sunrise Wednesday morning the whistle from the ponderous engine sounded a signal for the grand onslaught upon the sea of yellow grain. Ten headers and 36 header-wagons and an abundance of willing hands moved simultaneously with the machinery, and naught could be heard but the hum of the massive separator and the rattle and noise necessary among so many men, mules and headers. Four spouts poured out a continuous stream of grain. Four men attended to the sacks and four men did the sewing . . . At sunset the official count of sacks was made, the total being 6,183 bushels of wheat cut, threshed, and garnered from sun to sun. This showing we believe, is unprecedented in the annals of farming in the civilized world.

On West Coast farms, the use of steam engines had been limited because of the high cost of wood and coal fuel. With this in mind, H.W. Rice of Hayward, California, began as early as 1863 to manufacture straw-burning engines, which were fueled from the harvest leftovers. Others engaged in this manufacture included Joseph Enright of San Jose, J.L. Heald of Vallejo, and the Brown Brothers of Salinas. (10)

By the 1880s, there were 200 companies manufacturing farm steam engines in the United States, 192 of them east of the Rocky Mountains. (11) Although it would be expected that the first successful self-propelled steam engines would be developed in this region of heavy industry, it was California's Best and Holt manufacturing companies that achieved this important milestone. Whether it was the result of individual inventive genius, the pressures of large-scale farming, labor shortages, the high cost of horses, or the liberal cast of the western mind, the Far West was in the vanguard of technological innovation, first with the combined harvester and then with the steam traction engine.

It was clear to both Benjamin Holt and Daniel Best that plowing put more strain on horses than any other type of farm work. In addition, plowing by horse was slow work. Covering one section (one square mile) of land with a single 12-inch plowshare required a team to travel 5,280 miles. (12)

Not surprisingly, many attempts to build self-propelled steam plowing engines had been made by men such as Pennsylvanians Obed Hussey and Joseph W. Fawkes in the 1850s and Californians Philander H. Standish and Riley R. Doan in the 1860s. The results, however, had been meager. The C. & G. Cooper Company of Mount Vernon, Ohio, began converting portable threshing engines into self-propelled machines in the 1870s, but this was merely to allow the engines to be moved from one field to another for running power belts, not for pulling plows. Steam- powered plowing seemed feasible and was eagerly awaited by farmers, but the technology remained elusive.

In building the first farm steam engines, both Daniel Best and Benjamin Holt profited from the earlier work of George S. Berry of Visalia, California, who in 1886 constructed the first successful self-propelled combined harvester. (Remarkably, his innovation was constructed almost identically to ones in use today.) Seeking to speed up harvesting, George Berry and his brother William had converted a portable 25-horsepower Mitchell-Fischer steam engine, built in Oakland, into a self-propelled model. According to the "Visalia Times" of July 3, 1886, Berry's engine was mounted between the header and separator with drive wheels mounted in front of the steam boiler. As the straw came from the separator, it was pushed back onto a rack, after which a fireman pitched it into the fire-box. This was the first straw-burning combine in the United States.

In addition, steam was piped from the traction engine boiler to an auxiliary engine bolted to the combine. This flexible pipe became the first "power take-off" attachment used in American agriculture, an invention which became significant in later Holt patent suits. One appreciative observer commented on the beauty of seeing grain being cut on one side of the machine and chaff flying out the other, with sacks of grains being loaded in the middle on three chained-together wagons pulled by 16 horses. (13)

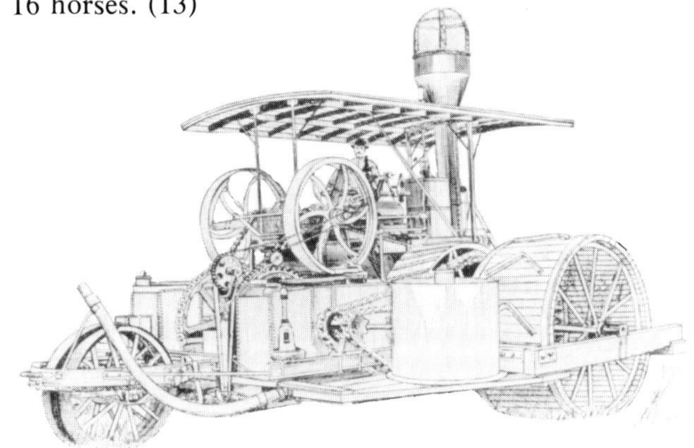

This Holt steam traction engine built in the 1890s was equipped with auxiliary wheels designed for better traction. Drawing in 1898 Holt y Company catalog.

Drawing in 1898 Holt Company catalog.

Holt steam traction engine pulling a steam driven combined harvester. Steam was carried from the main engine boiler to the engine on the harvester by means of pipes equipped with swivel joints. Photo about 1900.

Jacob Price of San Leandro, himself in the process of constructing a steam plowing engine, watched Berry's harveter in 1887, and wrote to the "Pacific Rural Press" that its performance exceeded his expectations. Thrilled to see the immense machine move off through a vast grain field with no horses within a mile, he commented in particular on its labor- saving aspects:

> When ready to start work, Mr. Berry asked me to take a seat by him on the tool box, which I did. Over our heads was stretched a large canvas awning some 20 feet square that protected all under it. At the word of the engineer, Mr. Berry's brother turned on the steam, and away we went cutting a swath 22 feet wide just as easy as winking. Just ahead sat a lazy fellow on a comfortable spring seat, twisting a wheel like a car brake to the right or left, guiding the ponderous machine to an inch. A fireman leisurely poked straw into the furnace to keep up steam. His hat hung nearby as if in a room, while his dog slept at his feet. In front of me was an engineer doing nothing at all, but alert and ready if required. On the other side of the boiler was a dusty looking fellow with an oil can in hand squirting in a little here and a little there, looking mighty wise and pretending to work.
>
> On a separate platform on the outside of the separator was a sack sewer comfortably seated in the shade sewing sacks and dumping them off on the ground at intervals. Altogether there were six or seven men who looked to me like they had a soft thing, even though they were putting in sacks ready for market 30 or 40 acres of wheat a day. (14)

In this same year, Jacob Price built a steam road roller in his shed in San Leandro, and used it to crush rock on the streets of San Jose. He then joined the J.I. Case Threshing Machine Company of Racine, Wisconsin, which built several of his nine-ton plowing steam engines. In June 1890, Price boasted in a letter to the "Farm Implement News" in Chicago that he had invented the first successful steam traction engine in the United States. (15)

Reflecting the highly competitive environment surrounding these inventors, De Lafayette Remington, scion of the famous rifle-making family, ridiculed Jacob Price's claims and insisted that he, Remington, held this honor. Having constructed a roughly made engine in Woodburn, Oregon, in 1885, he built an improved model called "Rough and Ready" which was patented in 1888 and taken to San Leandro where it performed before large crowds. Shortly thereafter, he sold his patent rights to Daniel Best who made some improvements before marketing the engines in 1889. In 1890, Best advertised them in the "Pacific Rural Press" as "Remington Traction Engines with Daniel Best's Combined Harvester". The words, "Daniel Best, San Leandro" appeared on the water tank of the engine. (16)

These Best engines were radically innovative in design. Instead of the conventional horizontal locomotive-type boilers, they featured a vertical boiler four feet in diameter containing 160 2-inch flues. Both the boiler and firebox rested between the drive wheels on heavy frame. The steam activated two 9-by-9 inch cylinders affixed to the frame in front of the boiler. Whereas traction engines made in the East had two front wheels, Best's tricycle machine had only one front wheel, which revolved in a turntable steering mechanism. The main drive wheels were 8 feet in diameter with a broad 26-inch face, making them ideal for crossing rough terrain. These wheels had a cog rim bolted to the inside of the outer rims, thus putting the power thrust on the outer rim of the drive wheel rather than on the axle, a design which Daniel Best claimed provided more power and better

A Holt harvesting outfit which carried a crew of five men: an engineer, a tiller operator for the header cutting bar, a separator man on the combined harvester and two sack sewers. Note the eight sacks on the dump platform. Photo about 1900.

This Holt harvesting outfit was equipped with a screen attachment called a smoke arrester. This attachment was designed to cover the smoke stack of the engine to prevent flying embers from the stack from starting fires in the fields. Photo about 1900.

Huge Best and Holt steam engines pulling in tandem in the logging camps of California. The Best engine in the lead was gear driven, while the Holt had a chain drive. Photo about 1907.

traction. Weighing 11 tons, the engine developed 40 horsepower with 100 pounds of steam pressure and 60 horsepower with 150 pounds of pressure. (17)

The first four Best engines were used in the Sacramento Valley in 1889 for plowing and pulling combines. Daniel Best's brother Henry operated one of these machines near Yuba City where he said they moved twice as fast as horses, doing the work of 75 mules and operating on the cost of the barley that the animals would consume. In 1890, the "Crocker Railroad Guide" admiringly described these steam engines as simply constructed with no superfluous pieces of wood or iron: "Trim, clean-limbed, muscular, steam-fanged, iron-ribbed giants ready to tear up the earth with twelve 12-inch plows running to the beam, at a tireless speed that would exhaust a hundred horses. (18)

In 1890, Benjamin Holt built "Betsy", his first steam traction engine, in Stockton. It was the first of Holt's series of 130 wheel-type engines which were to appear in the next 24 years. Holt No. 1 featured a 24-foot frame that supported a return-flue locomotive-type boiler which when filled with water weighed 45,000 pounds. A single 10-1/4-by-12 inch cylinder above the boiler drove a countershaft, which in turn carried the motive power to the drive wheels by means of heavy link chains. Believing chain drives to be superior to gears on combines, Benjamin Holt used the same system for his huge traction engines. This chain drive eliminated the transmission and differential gears, making for simplicity of construction. Friction arms controlled two very large flywheels, and the engine was reversible. These early engines burned 225 pounds of coal or 28 gallons of oil, and consumed 300 gallons of water per hour. Since the water tank held 675 gallons, it was necessary for a "tank man" to haul water to the machine while it was in operation. The horsepower ranged from 40 to 70 depending on the steam pressure. During the 1890s, Holt's steam traction engine sold for $4,500, with extension rims adding $500 to the total cost. (19)

In Stockton, "Betsy" went on to assume the sentimental role of factory mascot, being used for moving machinery and for loading flat cars on the railroad siding. Many young men learned to operate the other larger machinery by practicing on this engine.

Weighing in at 24 tons, Holt steam plowing engines were powerful brutes. Diary entries made in April 1893 by George L. Dickenson, secretary of the Holt Company, noted the progressive accomplishments of Traction Engine No. 3: "Started on Ashley's Ranch using 18 ten-inch plows . . . Now working with 24 ten-inch plows on Ashley's Ranch, going 1-1/2 miles per hour with three men—engineer, fireman and plow tender. Plowing four inches deep in heavy adobe. Success. (20) A photograph taken in the San Joaquin Valley in 1900 shows a Holt steam engine pulling 30 plow bottoms on five gangs equipped with grain seeders. In 1902 in the San Fernando Valley, another Holt outfit pulled a 96-foot hitch drawing plows, seeders and harrows over 40 acres a day, doing the work of 60 horses. (21)

Holt and Best steam engines were also used for hauling logs from timber sites to distant railroad terminals. Since

Moving the Julia Weber House in Stockton, California using three Holt steam traction engines. 1900 photo.

logging was done in rugged country, more power was needed to move the heavy logs to lumber mills, and the Holt Company built special timber trucks priced at $500, with a load capacity of 10 tons. Some had wheels 12 feet in diameter that could carry logs 6 feet across.

In California's Sierra Nevada near Emigrant Gap, trees were hauled over steep mountain roads by a steam logging truck pulling large steel wagons, five wagons to a train. Each wagon train had an engineer, a fireman, and a brakeman who set the brakes on each wagon by means of an iron rod which extended out from the wagon. The logs were unloaded by encircling them with chains and pulling them from the wagons with the timber truck, which was reportedly "almost as easy to handle as an automobile." (22) One account in the "Stockton Evening Mail" in 1903 claimed the freighting road engines, which were stronger than ordinary farm engines, could move timber or other freight at half the cost of animal power. (23)

Many of the new Holt and Best traction engines were used in road building projects. In the winter rainy season of 1890, the people of Sacramento hired one of Daniel Best's "Pathfinder" steam engines to haul gravel from a pit four miles from town to cover the sticky adobe street surfaces. Pulling a train of six large ore wagons night and day, the engine satisfactorily completed the job in a fraction of the time it would have taken horses. In Hawaii, another tireless engine plowed sugar cane fields in the day and drove a generator to provide electricity for factories at night in Honolulu. (24)

Holt steamer hauling lumber in Shasta County, California in 1902.

A Holt No. 37 engine used to pull down burnt out buildings after the April 18, 1906 San Francisco Earthquake.

A Holt steamer used to lay cable for the cable car system in San Francisco, California around 1900.

A bridge accident with a Holt steam engine. Many road bridges built for transport by horses were inadequate for handling traffic by heavy steam traction engines. The use of steam engines led to an improvement in bridge building across the United States. Photo around 1900.

Steam leviathans became known for pulling combined harvesters on the large grain ranches in the Pacific Coast states. (Their use, in fact, was limited to large scale operations, leaving a major portion of grain harvesting to horse-drawn combines and stationary outfits.) Awesome in appearance, some Holt steam harvesting units were 65 feet long, 64 feet wide, and 12 feet high. Their smoke stacks belched smoke from a height of 18 feet, and their two large flywheels churned the air like whirling dervishes. In 1893, some Holt combines had 34-inch-long threshing cylinders, 50-inch-wide separator units, and a 50-foot-long sicklebar on the header. (In the Midwest at the same time, grainbinders were limited to a swath of 12 feet.) The cost of the engine was approximately $5,000, the combine about $2,500. (25)

During the 1890s, the Best machines were advertised as "Monarchs of the Field," commanding the strength of 100 horses and able to harvest over 100 acres a day. A photograph on the Moreing Brothers' Ranch in the Sacramento River Valley in 1917 shows six Holt steam combines harvesting a field of 25,000 acres of wheat, a spectacle of giganticism in farm machinery never surpassed in the United States and remembered by all those who saw them at work.

One operator of a Holt steam harvester near Delano, California, in 1908 recalled that his outfit cut a 36-foot swath of wheat, moving at a rate of 4 or 5 miles an hour. His crew included an engineer, fireman, header tender, two sack sewers, a separator tender, two water bucks, one roustabout, one woman cook, and the boss. Between mid-May and September, they harvested 5,000 acres at a fee of $1.25 per acre, furnishing everything but the sacks, and they moved quickly from one ranch to the next:

> We tied the feed rack, oil tank and cookhouse behind the harvester, and away we went full speed. The two water wagons were each pulled by four horses; sometimes they would lead and sometimes they were behind. We never uncoupled the 36-foot header all season, but if any barbed wire fences were in the way, some of our crew pulled the posts and cut the wires in advance. When we pulled into another ranch, we cut off the feed rack, oil tank, and cookhouse, and we were harvesting wheat about as quick as it takes to tell it. (26)

With no morning dew to delay them, the men who slept in the stubble fields near the harvester or cookhouse rose at daylight, rolled their blankets, ate breakfast, oiled up, and were running across the field by 6:00 a.m.

Always looking for ways to improve the operation of Holt company farm equipment, Benjamin and his brother Charles followed one of their combines to Australia in the fall of 1894 to see it in action, as well as to scout out possible new markets. On January 19, 1895, Charles Holt reported to the company back in Stockton that the harvester gave unbounded satisfaction in Australia, making enormous savings of six bushels of

Plowing in South Africa with a Holt steam traction engine and three gang plows.

grain per acre over the stripper in common use and costing two shillings per acre to operate rather than 10. As a warning, he added that Benjamin did not want word about the excellent prospects for business in Australia to leak out to his competitors, so all glowing reports of the Holt machine's success should be strictly avoided. Benjamin and Charles were gone two months on their trip, arriving back in Stockton via Memphis and New Orleans some six weeks behind schedule. (27)

Occasionally Holt went to watch his steam combines at work in the fields. In 1904, he and his salesman Dan Gilmore arrived at a farm near Grimes, California. The young operator remembered that Ben Holt climbed up on the big steamer and opened the firebox for a look. "Young man," said Holt, "You have too much steam blast on. Cut that down and you will get better results." The farmer did, and his machine's performance improved. Dan Gilmore and Ben Holt made several more trips to Grimes that year to check up on three rigs. (28)

Operating these huge harvesters was no child's play. A glove caught in a flywheel could jerk an arm from its socket; 300 degree steam could inflict terrible burns. Exposed pulleys, belts, and shafts required the utmost care to be taken by the crew. On occasion, the heavy machines broke through wooden country bridges, endangering the lives of the crew. Repairs were often made late at night by the dim glow of a kerosene lantern.

Ripe grain was dry and highly inflammable, and fires routinely threatened the crews of steam threshing outfits. The "Pacific Rural Press" observed that steam driven combines worked where the temperatures reached 102 degrees: "If a match was touched to one of the dry fields, all creation would be on fire, so dewless are the nights, so fiercely bright the days."(29) Into these arid arenas moved threshing engines carrying a firebox full of flames and red-hot ashes. Although the machines' exhaust stacks were covered with screen spark arresters, flying embers still escaped into the air. When straw was burned for fuel and the engine was barking loudly under a heavy load, sparks flew as if from a blacksmith's anvil. After sundown these straw burners offered a display of fireworks, and some farmers switched to coal or oil to reduce this hazard. (30)

Every summer, farm journals in California ran articles with alarming headlines such as, "Fire! Fire!," "Beware of Sparks," and "State One Great Tinderbox." The "Pacific Rural Press" routinely reported incidents—a fire near Modesto which burned the separator and a stack of a grain amounting to a loss of $8,000, others which destroyed threshing rigs, wagons, farm buildings and grain fields, and one which devoured everything in its path until it reached the Sacramento River five miles away. Poet Joaquin Miller, a delegate to the American Forestry Congress in 1887, urgently spoke of the need to preserve California, the new Eden, "from sword, flood and flame," adding that fires had become so numerous that farmers could no longer secure fire insurance on their crops." (31)

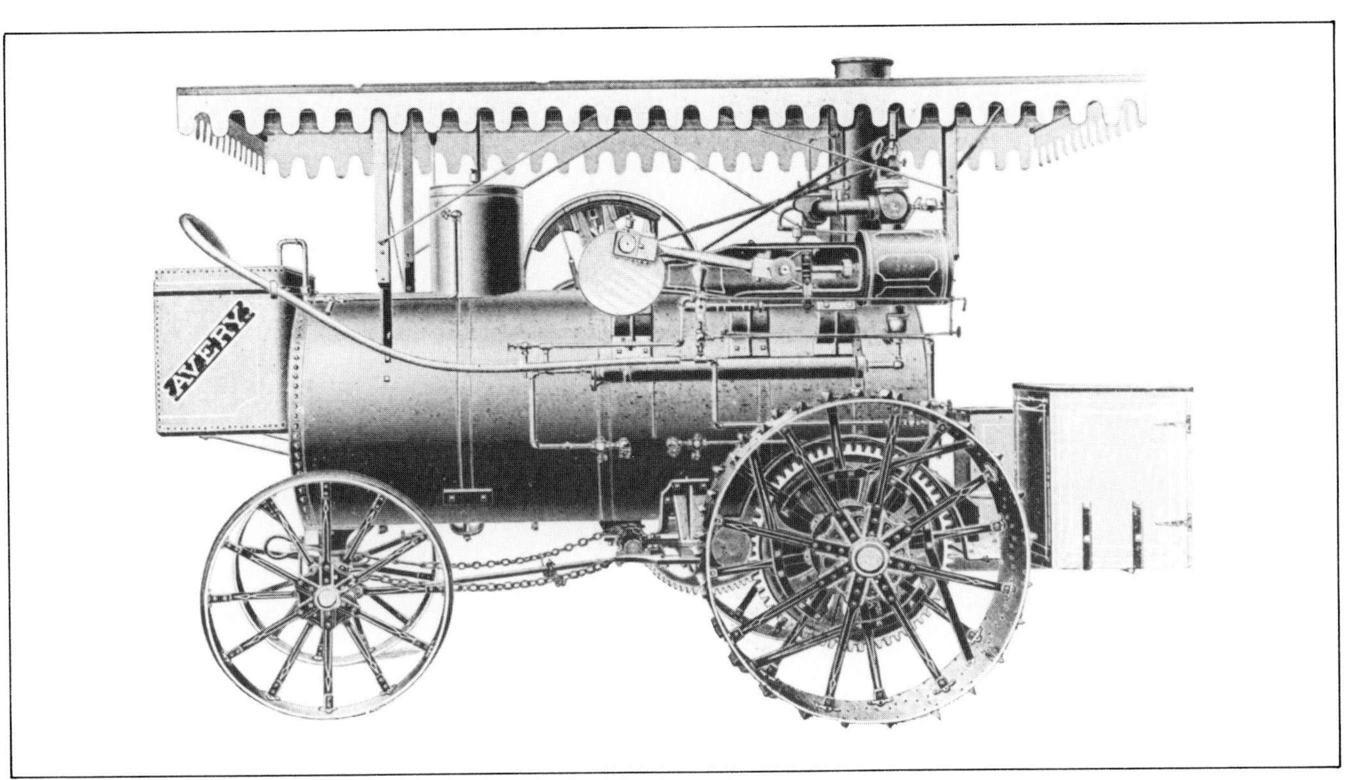

An Avery steam traction engine built in Peoria, Illinois, 1900. Drawing in "American Thresherman," March 1901.

A Reeves double cylinder steam traction engine manufactured in Columbus, Indiana in 1900. This company later produced a successful plowing engine. Drawing in "American Thresherman," January 1900.

Huge Case steam road locomotive. Regarded as the largest steam traction engine ever built in the United States. The J.I. Case Company built only three of them. Note that a man's head scarcely reached above the axle.

In addition to sparks from the engine, careless workers added to the fire hazard, and many threshermen begrudgingly switched from smoking cigarettes to chewing tobacco or snuff. In Oregon, Washington, and Idaho, fires were caused by a heavy growth of wheat smut, a fungus which formed a fine black powder during threshing and sometimes exploded from spontaneous combustion.

Methods of fire prevention varied. Some farm engineers carried an open pan of water under the firebox to catch hot coals that fell from the fire chamber. When fires started, some operators tried to control them by running the engine wheels over the flames. One driver recalled an experience near Williams, California:

> We cut the engine loose from the harvester, threw off the governor belt to race the engine, then ran the wide wheels right on top of the fire as a ten foot hose showered water over the wheels. Once we put out four miles of stubble fire using the 600 gallons of water in the engine—we used the outfit as a fire engine. (32)

Whenever fires were spotted, crew families and neighbors dropped their work and hurried to the area with barrels of water in which to soak burlap sacks used for beating out the fire.

Another hazard of steam engines was the danger of explosion. Under careful management, high-pressure boilers were relatively safe, but the danger was never completely eliminated, and engineers had to keep close watch on the steam gauge at all times. Newspaper accounts of threshing engine explosions abounded. One killed five men, and another near Rio Vista, California, in 1876 hurled a fireman 75 yards and scalded him from head to foot. (33) Another explosion drove an engine wheel through a barn wall, burned the separator, and destroyed a Methodist church and a nearby saloon in one great blast. In North Dakota in 1898, explosions killed 27 people and caused property damage of $118,000. (34)

Occasionally, engines exploded because engineers let the water get too low, thus exposing the crown sheet to the direct blaze of the fire. When these plates became red hot and the water pump injected a stream of cold water into this fiery cauldron, an explosion could occur. Some foolhardy engineers fired boilers which had no steam gauges, or they readjusted the safety valves contrary to manufacturer's instructions. Others worked the fields with rusty, unstable machines which no reputable engineer would have touched.

By 1900, complaints about boiler explosions led manufacturers to build boilers four times stronger than the earlier models. In addition, state laws were passed

Holt steam engine on Robert's Island in the Delta of the San Joaquin River Valley near Stockton, California. The wooden barrel wheel in front and the extension wheels denote efforts to keep the machine from sinking into the soft soil.

which defined specifications for boiler construction and required the licensing of engineers operating steam engines. By 1916, when the American Society of Mechanical Engineers adopted a code of standards for all engineers, however, steam engines were already losing their popularity in California and the Midwest and East.

It was not the hazards alone which caused steam engines to pass from the agricultural scene. Primary was the problem of the weight of the engine, which plagued both the manufacturer and the thresherman in the field. Although powerful, steam traction engines carried about 1,000 pounds of weight for every horsepower delivered to the drawbar (about ten times the weight ratio in modern tractors). As a result, although the ponderous steam engines had plenty of power to turn their drive wheels, they mired down in soft ground or sandy soil and lacked adequate footing to extricate themselves from mud holes. Once an engine was buried to the axles, it could take three days of working timber and jack-screws to extricate the machine.

Although much of the farmland in the Central Valley could provide reasonably solid traction for these steam engines, the Sacramento-San Joaquin River Delta region created serious problems. Draining one-third of California, the Delta comprised 535,000 acres stretching 48 miles from Stockton north to Sacramento, half of which lie below sea level. Extremely fertile, this peat acreage was created by the decomposition of tule plants and alluvial deposits from river systems. (35)

For many years, farmers had known that the Delta lands were as rich as the Valley of the Nile, but no one had found a way to mechanize farming on the soggy peat soil. Prior to 1870, hundreds of miles of levees had been built by Chinese laborers, who cut slabs of tule sod with knives and moved the earth to the levees by wheelbarrows. Later, horse-drawn scrapers and steam-operated dredges completed the levee work. But the spongy soil remained so soft that it failed to support horses, even when they were shod with special "tule shoes" measuring one foot in diameter. Similarly, wagon wheels with special 10-inch-wide rims sank into the peat lands. (36)

Keenly aware of the economic potential of the Delta, Benjamin Holt and Daniel Best experimented with steam engines equipped with wider drive wheels during the years from 1889 to 1908. Since Benjamin and Charles Holt owned peat land on Roberts Island near Stockton, they were eager to mechanize their farming operation. To get more traction for their steam engines, they added to each 6-foot-wide drive wheel several extension wheels, until each rear drive wheel was 18 feet wide. Because the engine itself was 10 feet wide, roads and gates to fields needed to be 46 feet wide to accommodate this goliath. In front of the machines, they placed an additional barrel-like wheel to help support the machine and a superstructure of wooden timbers to prevent breakage of the main axles. (37)

Although these monstrosities did some work in the Delta lands, they were inefficient to operate and costly to manufacture. They could not turn around easily at the end of a field. By increasing the size of the wheels the machines' weight also increased, creating a self-defeating mechanical dilemma. The problems of weight, size, fire danger, and maneuverability were not resolved until further advances in agricultural technology pioneered by Benjamin Holt and others set the stage for a new era of streamlined, widely-adaptable farm machinery.

Steam-powered traction engines had revolutionized farm work in their era, and the engines developed by Daniel Best and Ben Holt in 1889 and 1890 led the advance. They did not, however, dominate the total market after the early years of manufacture. By 1910, Holt and Best machines produced 6,500 of the total 3,600,000 steam horsepower, or two percent of machines available for work on U.S. farms. Since introducing the machines to the market, they had manufactured a total of approximately 300 steam traction engines, while J. I. Case, Huber, and Minneapolis companies produced about 71,000 machines. (38)

As for the pioneering men who had worked the powerful but cumbersome steam traction engines, some of them moved to the new era with reluctance. Farmers joked that a retired thresherman had to be blindfolded whenever a new threshing rig passed his farm to prevent him from dashing to town and mortgaging his home to buy another outfit. Victims of "thrashing fever" reportedly had "thrashin' in their blood", and steam men long reacted to the sensations associated with their job—the smell of live steam passing over an oily boiler, the heart beat of the injector, the barks from the stack, the gyrating governor balls, the sight of a combine or threshing machine "hogging" through heavy grain sending up a cloud of dust, and the blast of the whistle at day's end. When one old steamman was asked in 1920 why he remained on the job year after year, he wiped his hands on his greasy overalls, shifted his cud of tobacco, expectorated copiously, and reflected:

> I recon I have sworn off this durn thrashin business a hundred times. Every year I say, well this is my last, but it ain't. It ain't money. Lordy you don't get rich, hustles me to break even lots of times. I recon I just naturally have a hankering to be oily and greasy and covered with dust and be jawed at and work all day and half the night. I swear off, and durn it if I ain't crazy as a kid just as the threshing season starts. (39)

Other threshermen welcomed the new advances in farm machinery. One recalled:

> The most dreadful days of my life have been spent with threshing machinery. I have arose at three o'clock in the morning to start a fire in the dearest engine this side of the Mississippi. The boys would bring me breakfast of soda biscuit and sow belly. I have walked majestic and solemn at midnight and drank water that would make an engine foam in 15 minutes. I have shouted "Come on boys" until I have almost lost faith in mankind and fell in love with the hired girl because she had on a clean apron. (40)

While many steam threshers were used for years after more advanced machines were available, most of them eventually found their way to junk yards. World War II scrap iron drives claimed the last remnants, and museums of technology and industry overlooked them for years. Today, not a single Holt steam traction engine has survived.

Collectors of farm machines, however, now restore and display many ancient pieces of farm machines at annual "Threshing Bees" and "Pioneer Days." A quarter of a million people flock annually to Mount Pleasant, Iowa, to see old steam engines perform at a five-day threshing festival—a testimony of their nostalgic power.

Holt steam traction engine with extension wheels designed to give better traction. Scene in San Joaquin Delta near Stockton. Timbers were used to prevent breaking of the engine axles revealing a need for track-type tractors.

Chapter 6

BENJAMIN HOLT'S FAMILY AND ASSOCIATES

The growth of the Holt enterprises was not limited to the work of Benjamin Holt alone. He received solid support from family members, company mechanics, factory workers, office staff, lawyers and salesmen. All played a part in making the Holt operation a success.

For its day, the company's business structure and practices were relatively sophisticated. In 1883, when Benjamin moved to California, his father, William K. Holt, and his brothers owned three wheel and carriage companies; the original family business in Concord, New Hampshire, one in San Francisco and one in Stockton. In that same year, Benjamin's brother William H. Holt retired and sold his interest in the Stockton Wheel Company for $65,000 to his three younger brothers, Charles, A. Frank, and Benjamin.(1) This left Charles heading the San Francisco branch, Benjamin overseeing the Stockton Plant and A. Frank directing the Concord factory. When A. Frank died in 1889, Charles and Benjamin purchased his share of the Stockton Wheel Company, thus becoming its sole owners. They also bought A. Frank's interest in the Concord company, and Charles assumed its presidency in 1889.

As the business in Stockton expanded, Charles and Benjamin reorganized the business in 1892 and incorporated the Holt Manufacturing Company with capital stock valued at $200,000. At the same time, they incorporated the Holt Bros. of San Francisco with the same amount of capital stock. Charles remained president of the San Francisco firm, with Benjamin as vice-president, and in the Stockton company, their positions were reversed. Both held these posts until their deaths. Managing three corporations on two coasts, the two brothers had established a business organization similar to a modern conglomerate.(2)

Benjamin and Charles' business affairs became even more complex when the Holt Manufacturing Company in Stockton expanded by purchasing other corporations. In 1895, it acquired the Matteson and Williamson Company, one of the oldest manufacturers of farm machinery in the state. In 1901, the Holts bought the Houser and Haines Manufacturing Company, a Stockton plant covering two city blocks. Charles remained president of this corporation until his death in 1905. On October 8, 1908, Holt Manufacturing bought

A Benjamin Holt picture taken about 1911. Former employees of the Holt Manufacturing Company consider this picture a very representative portrait. He often wore this old overcoat around the drafty plant.

out its strongest rival, the Best Manufacturing Company of San Leandro, California. By purchase and merger, the Holt family then came to hold a virtual monopoly on the production of combined harvesters in the United States, as well as a strong position in other lines of farm machinery.

The proliferating Holt interests soon formed a maze of interlocking corporate entities. A branch factory was acquried in Walla Walla, Washington, in 1902 and a branch office on Spokane in 1908. Holt incorporated the Aurora Engine Company in Stockton in 1906 with its own officers and stockholders to manufacture gasoline motors. A Northern Holt Company was established in Minneapolis in 1909 to handle sales in the northern plains, and another factory opened in Peoria, Illinois, in 1910. In 1914, the Canadian Holt Company, Ltd., was set up in Calgary, Alberta.(3)

Remarkably, most business operations for these companies were handled by a small office staff in Stockton. George L. Dickenson, a shrewd man with precise habits, served as Benjamin Holt's secretary and counselor. Leaving in 1910 for similar duties with the new Holt Caterpillar Company in Peoria, Dickenson was followed by Ollie H. Eccleston, who served as secretary until 1925.

Charles L. Neumiller, a rotund lawyer and one of Benjamin Holt's most trusted advisers, successfully consolidated the various Holt operations into one major corporation in 1913 at a critical time in the company's growth. (Benjamin Holt owned 75 percent of the new corporation's stock). Neumiller, a tough-minded pragmatist who drove a hard bargain, especially during the merger with C. L. Best Tractor Company in 1925, kept family members working smoothly together.(4)

In the main office, George H. Cowie acted as treasurer and sometimes took charge of buying and shipping. Thomas H. Luke managed the parts department where he displayed a remarkable memory for the specifications of thousands of machinery parts. Machine shop foremen Russell S. Springer rose through the ranks to become an assistant to one of the vice-presidents, a factory manager, and vice-president in 1913.(5)

As the years passed, Benjamin and Charles Holt received assistance from the next generation of Holts. Benjamin's son, William K., worked closely with his father in the factory and accompanied him on many business trips. Later, William K. organized a Caterpillar dealership in Mexico City and San Antonio, Texas.

Charles' son, Charles Parker or C. Parker as he was always called, gained "hands-on" experience in the Stockton plant as a draftsman, boilermaker, and steam fitter to add to his advanced engineering training. When Charles died in 1905, C. Parker assumed the duties of a vice-president and director of the Holt Manufacturing Company. In 1906, he sold his inherited interest in the corporation to Benjamin in exchange for Benjamin's interest in the San Francisco company. When Benjamin asked C. Parker to return to Stockton in 1909, he reinvested there, assumed a vice-presidency for the second time, and later became treasurer. While in charge of the company's exports, C. Parker fully used his extensive practical and theoretical education, as well as his urbanity and social skills, to bring Holt Caterpillar machines to the attention of government officials and businessmen around the world.

Pliny E. Holt, the son of Benjamin's older brother, William H., graduated from engineering school and joined the Holt Manufacturing Company in Stockton in 1896. An excellent and imaginative engineer, he received 38 patents and collaborated with other company engineers on approximately 60 other patents, most of which applied to tractors, combined harvesters, and artillery pieces for the Ordnance Department of the United States Army.(6) In 1905, he organized and became president of the Aurora Engine Company and served as treasurer of the subsidiary Houser-Haines Company.

Pliny E. Holt's most outstanding achievements were his work with his Uncle Benjamin to develop the Caterpillar tractor, the establishment of the Holt factory in Peoria in 1909, and the invention of various military vehicles for the United States Army. Pliny also served as vice-president and director of the Stockton firm from 1905 to 1925 and as a director of the Caterpillar Tractor Co. in Peoria until his death in 1934.

Pliny's brother, Ben C. Holt, joined the parent company in 1907, took over as manager of a branch factory first at Walla Walla, Washington, and then in Spokane, and served as a vice-president and director of the Stockton corporation. Despite a speech impediment, he had the reputation for being one of the best salesmen on the Pacific Coast. (He claimed that farmers to whom he was selling harvesters felt sorry for him, and before they realized it, he had their signature on the order blank).

Dan N. Gilmore became a super salesman, earning as much as $98,000 in commissions in one year.(7) He was so good in selling tractors that his friends in jest claimed he could sell rowboats in the Sahara. His skills led to a promotion to posts of managerial responsibilities within the company. It was people of this type that contributed to the success of the Holt Manufacturing Company.

Benjamin Holt was one of the first Stockton citizens to take a passenger flight in an airplane. His son, William K. Holt, later said that his father was not afraid of anything.

Holt Steam Traction Engine No. 73 standing by the Holt factory in Stockton. All Holt steam engines were designed with return flue boilers. In case of fire the hose in front running to the boiler could be used to spray water on the flames. Some engineers also used the drive wheels to run over the fire line in the fields.

The header section of this Holt harvester was hinged so that it could be trailed on the road more easily. Photo about 1898.

Chapter 7

BENJAMIN HOLT AND THE TRACK-TYPE TRACTOR

Although Benjamin Holt was widely known in 1900 for manufacturing mammoth steam traction engines and combined harvesters, it was the track-laying tractor that was to be his most innovative and notable achievement. More than any other person, Holt was responsible for developing the first successful crawler farm tractors used in American agriculture, a dramatic technological breakthrough capping centuries of experimentation.

By improving the efficiency of the conventional wheel, the track-type tractor furthered man's ability to apply mechanical power to challenging tasks. By using a set of chain-link tracks for wheels normally used to support self-propelled machines, the track-type tractor carries, lays, and picks up its own tracks as it moves across the ground. Because these tracks distribute the weight of a machine over a much larger surface area than do ordinary wheels, which touch the ground at only one point, the track-laying engine is able to traverse mud, sand, and snow and to climb a steep incline in the same way that skis enable a person to move on top of snow.

Although widely known today, the track-laying tractor has obscure origins. Egyptians probably put rollers under heavy stones to build the pyramids, and medieval man may have used the same technique to move large catapult weapons. The crawler-type tread concept was used by a Frenchman named M. D'Hermand of Paris in 1713, who made a drawing of a treadmill device consisting of 26 rollers moving as a belt. Pulled by goats, this machine was designed to carry children in city parks. In Great Britain in 1770, Richard L. Edgeworth applied steam power to a small vehicle like that as suggested by M. D'Hermand, but Edgeworth left no drawings or models. He mentioned in his memoirs that the Frenchman had already preempted his idea.(1)

Later an Englishman, Sir George Cayley received a patent in 1825 for drawings for an apparatus which placed an endless chain belt over the two wheels on each side of a wagon. This design would prevent the wagon from sinking in soft terrain, but the inventor failed to provide a mechanism for steering the wagon when in motion.(2)

In 1713 M. D'Hermand of Paris, France built a cart mounted on rollers which formed an endless track. This may have been the first model of the principle of a track-type vehicle.

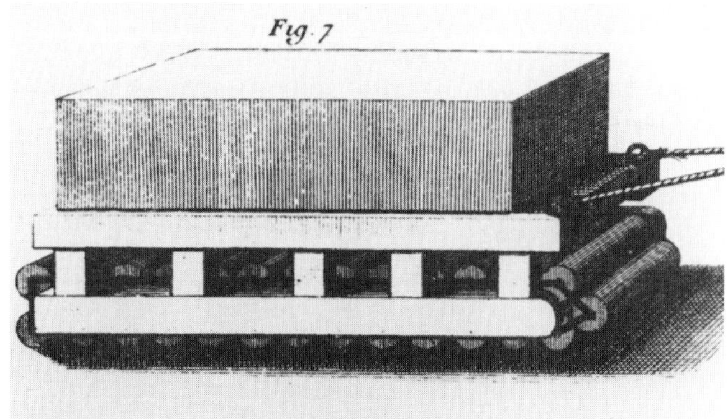

George Cayley's patent number 5260, December 5, 1825. The English patent stated that George lived in Brompton, York County. His patent was entitled "Cayley's apparatus to be attatched to carriages." Copy of the original patent reprinted by Love and Malcomson Ltd., March 25, 1916.

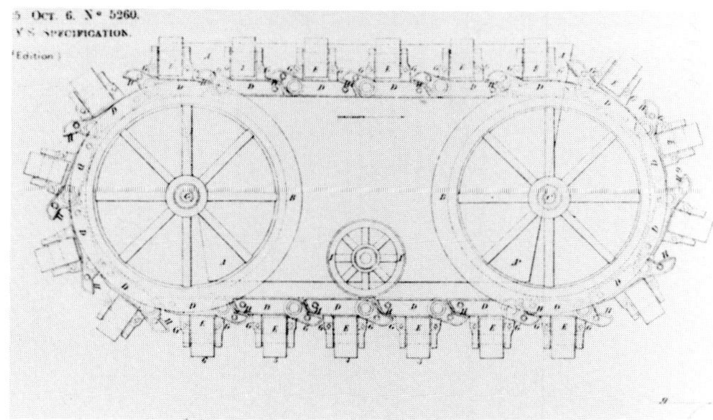

In 1832, a British textile manufacturer, John Heathcoate, built a 30-ton steam plowing engine which rested on two 7-foot-wide belts. These tracks were intended to prevent the steam engine from sinking into marsh land. Attracting considerable attention, the engine was heralded as "a remarkable contribution to science and to the wealth of the country," but optimism was shattered when the monster sank and almost disappeared in a swamp. Heathcoate lost $12,000 in the venture.(3)

Because the belief in the use of the wheel remained so ingrained, many would-be innovators continued to focus on the idea of attatching things to ordinary wheels in order to give them more traction. Influenced by the tradition of the horse, some inventors developed wheels with various types of feet or shoes to push against the ground. James Boydell of England invented an "elephant foot" steam engine in 1846 which featured six wooden paddles bolted to the wheels of the machine. These flat shoes were put down and picked up as the wheels revolved, making a loud clattering sound. An American who observed the engine at work admiringly noted, "It walked over the earth like some huge animal, puffing and snorting, and taking its six plows along with no apparent consciousness of effort."(4) Although these wheels provided more ground-bearing surface, the accessories contained many moving parts that created friction and required frequent repairs. In addition, the moving parts failed to function when jammed up with mud, sand, or frozen ground. Improving on the simple wheel proved to be more difficult than imagined.

Some inventors hit on a new approach but did not have the money to see their idea through to production. In 1858, Warren P. Miller of Marysville, California, constructed a track-type steam plow which he demonstrated at the State Fair. Although Miller received a gold medal and cash prize for building a machine suitable for farming, only this one machine was ever constructed. Drawings of the machine found in patent literature indicate that it incorporated all the basic design features used in modern crawler tractors, including independently driven tracks for better steering. (In fact, this engine was used as an example of "a prior art" in lawsuits involving the patents for track-type tractors.)(5)

After the Civil war, numerous individuals experimented with track- laying machines with limited success. George Minnis used a steam plowing engine equipped with movable tracks for breaking sod near Ames, Iowa, which the United States Department of Agriculture praised in its annual report of 1869. In 1871, R.C. Parvin of Philadelphia designed a steam engine driven by crawler tracks and demonstrated it at the Illinois State Fair. He then organized a company in Chicago to produce the machine in quantity, but this venture failed.(6)

James Boydell of England built a steam engine equipped with platform wheels. The machine was driven in 1846, but the wooden timbers bolted to the wheels proved to be mechanically unsound. Apparently Boydell believed that the principle of snowshoes and skis could be applied to wheels.

However promising these machines seemed, all of them proved to be only experimental prototypes and none were produced commercially to meet the power needs of American agriculture or industry. Success in this venture had to wait until the twentieth century.

The first steam engines with crawler tracks manufactured and sold commercially were produced by Alvin O. Lombard of Penobscot County, Maine. A millwright, Lombard had observed the difficulties encountered in moving logs through deep snow in winter. Since horses proved inadequate for this task, he built a steam engine designed very much like a railroad locomotive and driven by a set of crawler tracks. A main frame supported the boiler and engine, while a supplementary frame held sprocket wheels for driving the link-belt chain treads. The weight of the machines rested on rollers running inside the endless chain tracks, while the front end rested on a sled with runners for easy use in the snow.(7)

Completed in mid-June 1900, this engine was given field tests for five weeks. It worked well on level ground, but since the frames were bolted solidly together, the machine lacked flexibility in crossing rough ground. To eliminate this problem, Lombard attatched the supplemental frame holding the tracks to a large shaft which pivoted in a journal box, thus permitting the tracks to accommodate irregular land surfaces. Lombard patented the design for this improved model on May 21, 1901. The patent document suggests that Lombard included the essential features of the Warren P. Miller engine built 43 years earlier, with the only significant innovation being the sled to replace the front wheel for use in the snow.(8)

A diary kept by Alvin Lombard's daughter documents that Lombard built three locomotive log haulers in 1901 and 1902. The first was sold to a logger in Waterville, Maine, in 1903; the second engine went to the Conneticut Valley Lumber Company in early 1904; and the third was shipped to the Eau Claire Mills and Supply Company in Wisconsin in May 1904.(9)

These early Lombard steam locomotives created a sensation in the lumbering regions of New England. Their endless tracks appeared to make possible the transporting of logs throughout the winter months. When one of these impressive outfits reached Colbrook, New Hampshire, in 1904, it was hitched to a number of sleds loaded with 40,000 feet of logs. Alvin Lombard recalled that "hundreds and thousands of people lined the road; they came from Boston and from the lumber mills in the woods, and ministers had to shut down the churches on Sunday so people could see the event." After this demonstration, an official of a lumber company purchased the engine on the spot with a $5,000 check.(10)

In 1904, the Eau Claire lumber firm established the Phoenix Company and secured a license to manufacture the logging locomotives, in return for agreeing to pay a $1,000 royalty to Lombard for each engine produced. Weighing 17 tons, moving on tracks 16 inches wide, and traveling up to five miles per hour, these Phoenix machines could pull 15 logging sleds loaded with 20,000 tons of logs. Between 1904 and 1915, the Phoenix Company sold 65 of these logging engines, while Lombard's company in Waterville, Maine, sold another 140. By 1915, over 200 Lombard track-laying logging engines had been manufactured and sold in the lumber regions of the northern United States.(11)

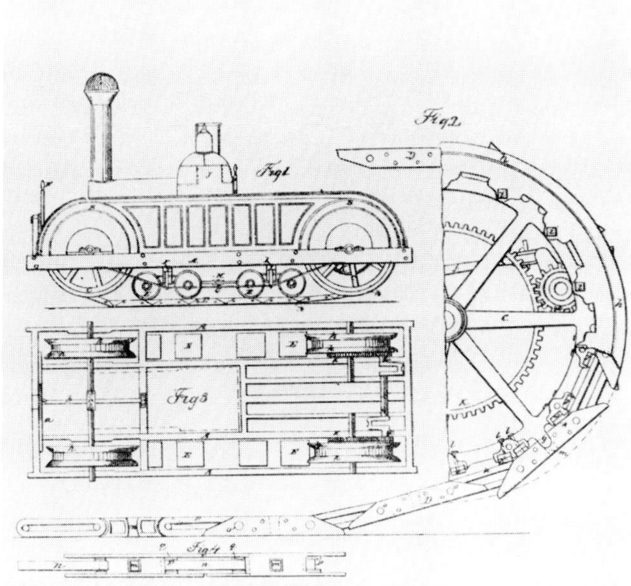

In 1858, Warren P. Miller of Marysville, California built a "steam wagon," the first crawler to be demonstrated at the California State Fair in Sacremento. Miller was awarded a medal and cash prize for building a machine that was suitable for agriculture. This machine had a design which included virtually all the basic principles in manufacturing track-type engines. Thus it became a prime example of a "prior art" in legal cases dealing with patent rights of various inventors.

An unusual feature of the Lombard log hauler was the placement of a steering mechanism in front of the locomotive boiler to guide the two sled runners. A steersman seated close to the smokestack in a special cab guided the engine by turning a large iron wheel. Since the locomotive had no brakes, hay was often scattered on downhill slopes to prevent the whole train from racing down the snow-covered trail out of control. A steersman recalled:

> Getting squashed was just one potential problem. The steersman had at least two other very real problems to contend with. Maine winters get mighty cold—cold and still— unless you're sitting in front of a Lombard going downhill at 20 miles per hour . . . You could be perspiring profusely, wrestling the wheel, while your nose and toes froze. Even worse, when the log hauler hit the flats and slowed to four miles per hour, the wood sparks and smoke often shot out the smoke stack into the steersman's shack and burned the wool mackinaw off his back as he was coughing, choking and wiping the frozen tears from his eyes.(12)

Although Lombard was the first to prove that crawler tracks could be manufactured and sold on a commercial basis, his steam-powered log haulers were limited to lumbering work. It was on the West Coast that the modern, all-purpose track-type tractor was developed by Benjamin Holt to meet the demands for power created by large-scale farming.

Even though advances in technology in the 1890s enabled California to produce 40 million bushels of wheat on almost three million acres of land, steam traction engines and horses for plowing and harvesting were far from satisfactory. Steam harvester outfits frequently required crews of seven men, making labor costs as high as on horse-drawn combine rigs. In addition, steam-powered units required a skilled engineer, while any farmer could handle teams of horses. Heavy steam engines also became easily mired in soft ground. These deficiencies were observed by Benjamin Holt, who turned his attention to perfecting and manufacturing the track-type tractor.

Holt had been concerned with the problem of engine weight and traction since constructing steam engines with 18-foot-wide drive wheels for use on the soft peat land of the San Joaquin Delta in the 1890s. Conventional wheels, he realized, simply had too little ground-bearing surface—it would require a 75-foot-high wheel, he calculated, to secure a 5-foot bearing surface, hardly a sound engineering prospect.(13) Reasoning that there must be a better engineering principle, he began toying with the idea of putting two wheels in tandem and connecting them with an endless chain-link belt to maximize the machine's traction in relation to its weight.

How did Holt reach this novel concept? Perhaps he remembered the tread-power machines used on farms in New Hampshire to provide belt power for grinding grain, sawing wood, and threshing grain. Tread-power

Lombard track-type steam engine utilizing chain drive principle.

One of the first successful track-type traction engines was assembled by Holt around 1905. The tracks were 9 feet long and were made of 3x4 inch wooden slats.

machines had been in use for a century, however, and it seems surprising that no one would have made this connection sooner. He may have gotten his idea from a ride to the top of New Hampshire's Mount Washington on a cog railroad. As he watched the cog wheel mesh with a slotted third rail and observed the wheels move over the smooth rails, he may have realized that a similar vehicle might be made that would in effect lay down and pick up its own rails, and thus not be confined to a fixed railroad track.(14)

Possibly it was the immediate problem of moving engines across the spongy Delta land that called for a novel solution. Holt knew that the family-owned land at Roberts Island could grow profitable crops if the work could be mechanized, and this economic pressure may have become its own incentive for the inventive-minded engineer.

Another theory holds that Benjamin Holt solved the traction dilemma by thorough research into the problem and possible solutions. Working like a scientist, this theory holds, he carefully examined what had already been done and then followed a painstaking program of personal investigation. It is true that he and his nephew Pliny visited Europe in 1903 to learn what had been accomplished with crawler tracks. Pliny later recalled, "We went all over the United States and European high spots to see what had been done with tracks. We saw crawlers in various stages of perfection in Illinois, Kansas and Maine."(15) With this research behind them perhaps the two men blended these mechanical ideas into what was to become the Holt track-type tractor.

Also the argument can be made that it was Benjamin Holt's innate creativity that led to this invention of the first successful farm crawler tractor. Perhaps it was Holt's genius that produced this technological breakthrough.

However, Alvin Lombard, the inventor of the track-laying logging locomotives, maintained that Benjamin Holt simply copied the track principle from Lombard's machines without making a cash settlement or paying royalties. Lombard testified to this opinion in court and stuck to it as long as he lived.

Regardless of the source of Holt's inventive inspiration it is apparent that he pragmatically plugged away at the problem of farming on the soft delta lands near Stockton, probably without thinking that his ideas would ultimately have worldwide consequences. With his eyes on the immediate goal, he instructed his mechanics in the fall of 1904 to remove the rear drive wheels on the 40 horsepower Holt Junior Road Engine Number 77 and replace them with a pair of tracks 9 feet long and 24 inches wide. The track shoes consisted of 2 by 4 inch wooden slats or plates bolted to an endless chain. These chains were driven by sprockets while the weight of the

An early experimental Holt track-type steam engine built around 1904.

engine moved on rollers journaled to a supplementary frame— a design very similar to that of Warren P. Miller's prototype built 46 years earlier.(16)

On Thanksgiving Day, November 24, 1904, Holt's steam engine was given its first field test, where it performed in superb fashion. As the machine returned to the factory it passed along Aurora Street where Benjamin Holt; John Shepard, a painter; and Charles Clements, a photographer; were standing. In passing the moving tracks created an optical illusion giving the impression that the top of the track was moving forward while track resting on the ground was standing still. As Clements observed this odd movement he exclaimed, "It crawls just like a caterpillar." Benjamin Holt, struck by this apt expression, replied, "Caterpillar it is. That's the name for it."(17) The name "Caterpillar" as a trademark was registered with the United States Patent Office in 1910 and has remnained the sole property of the corporation.

Sensing the importance of this event, Benjamin Holt spent much time in supervising the experimental work. George L. Dickenson made several entries in his diary stating that Ben Holt had been out to see the "Paddle Wheel" engine work on the Holt Ranch on Robert's Island. Other comments were, "March 29, 1905. Ben Holt down to Ranch this p.m. to see traction engine with mud turtle wheels work. He claims they are a great thing."(18)

Journalists soon noticed Holt's new machine. A front page story in "Farm Implement News" on May 18, 1905, "A Novel Traction Engine: Plowing Engine Fitted with Platform Wheels," reported enthusiastically:

> In the Roberts Island tract where a man could not walk without sinking to his knees and where tule-shoed horses could not be used . . . the new traction engine was operated without a perceptible impression in the ground. Hauling gang plows over the land, the machine accomplished the cultivation of the tract in a most satisfying manner . . . It is predicted that with the new device it will be possible to work any of the soft lands of the reclaimed districts and bring into cultivation thousands of acres of rich areas that are now unproductive.(19)

Russell S. Springer, a Holt Company machine shop foreman at the time, recalled that this first tractor's tracks were variously called platform wheels, mud turtle wheels, railroad wheels, tread mill wheels, paddle wheels, and "caterpillar" wheels. In 1935, Ben's nephew C. Parker wrote that:

> . . .the first "Caterpillar" wallowed around so successfully in the soft mud of Mormon Slough . . . that it was sent down to the Holt Ranch where it operated steadily for an entire winter. During this time a second steam tractor was fitted with the new track-type wheel and was sent down into a very sandy vineyard near Manteca. There it operated with very great success.(20)

After testing six different models of track-type steam engines in 1905, the Holt Company sold its first new machines to J.M. Jefferey in 1906. Jefferey's Golden Meadow Development Company of Lockport, Louisiana, was engaged in clearing Mississippi Delta land below New Orleans. This firm had previously purchased a large Holt wheel steam traction engine which could not do the work. The order for the new machine read:

> Ship Dec. 21, 1906. Engine ÷111. A52387. $5,500.00. Holt Brother's Improved Traction Engine. A Paddle Wheel Traction Engine with 9 1/2 by 12" cylinder. A 600 gallon water wagon and one plow cart . . . All payments to be made in Gold Coin, San Francisco or New York Exchange.(21)

Finally, Holt's years of experimentation with traction engines had paid off and produced a commercially viable machine. A new engineering solution now enabled powerful machines to traverse and work land that had never been farmed by mechanized vehicles. The ramifications of this technological breakthrough proved to be immense, probably more than Holt himself could have dreamed.?

Pliny Holt at the control of the first successful track-type traction engine. It was assembled on November 24, 1904 when workmen removed the rear wheels of steam engine number 77 and replaced them with a set of tracks.

Chapter 8

THE GASOLINE ENGINE ARRIVES

The early years of the twentieth century ushered in an age of power built around the advent of the internal combustion engine. The gasoline motor made possible the building of automobiles, trucks, tractors and airplanes.

For many years engineers had dreamed about gasoline motors which would deliver the same power as steam engines but weigh only half as much. Gasoline engine tractors, men like Ben Holt knew, would eliminate perplexing problems such as boiler explosions, fires, and the need for extra manpower to haul water and coal to the insatiable steam engines. The idea of a one-man tractor which could be operated by the average farmer rather than a trained "farm engineer" became a strong incentive for applying the internal combustion engine to the new crawler tractor vehicles. (1)

Virtually all of the early gasoline powered tractors were built by the manufacturers of farm steam engines rather than by companies producing gasoline-engine automobiles. As a result, the first tractors were not "Made in Detroit" but were born in such places as Racine, Wisconsin; Richmond, Indiana; Minneapolis, Minnesota; Charles City, Iowa; and Stockton and San Leandro, California.

Hart-Paar advertisement in "The Canadian Thresherman," March 1906.

Farm implement firms such as California's Holt and Best corporations already owned large factories, had good capital reserves, and boasted a long list of customers. They had branch managers, dealers, and road men to make sales, deliver machinery, and provide repair services. In addition, by the turn of the century most production problems in manufacturing heavy farm machinery had been solved. Qualified engineers were available, a machine-tool industry already existed, and corporation executives had gained experience in understanding managerial issues. When the decision was made to turn to gasoline engines, all that was required were changes in blueprints and the building and mounting of gasoline motors on existing chassis which had previously carried the boilers of steam traction engines. As a result, the conversion in farm machinery from steam to gasoline power occurred quickly, and the new industry was launched almost full grown.

Both Benjamin Holt and Daniel Best began building gasoline engines at an early date. As early as 1892, Best was providing gasoline motors for street cars on an eight-mile line running from San Jose to Alum Rock, the same year that Charles Duryea drove his first automobile in Massachusetts. Best tested a gas tractor in 1896, the same year that Henry Ford produced his first automobile, and on July 4 of that year the "San Leandro Reporter" described a tug-of-war between Best's gasoline tractor and a steam traction engine. When the engineers gave full throttle to the two engines chained back-to-back, the tractor pulled the steam engine around the block to the cheers of an enthusiastic crowd. (2)

"Do it yourself kit." Some early tractors were virtually homemade. In 1906 it was possible to buy a tractor chassis from the Temple Pump Company of Chicago, Illinois. The farmer could then attach any stationary gasoline tractor to this chassis and the tractor was now driven by a chain drive. The owner might use the stationary engine indoors during the winter months and then move it to the chassis in the spring for field work. The farmer could also buy the motor from the Temple Traction Company choosing a size from 20 to 45 horsepower.

Daniel Best was also among the first to build an automobile on the West Coast. In September 1898, he appeared on the streets of San Leandro in a motor carriage powered by a seven-horsepower engine which could go 20 miles per hour. The car, which had an electric ignition system and hard rubber tires, could carry eight passengers. In later years, Best recalled being smitten by "auto fever" and deciding to build one for himself:

> That machine was a work of art. It had the grace of a mud scow; it was a nine-day wonder that ran for eleven years. I constructed a second machine, and later gave it to my son. He in turn traded it for a piano. I think the piano man was cheated. I have often thought if I had stayed with automobile manufacturing I could have out-Forded Ford. Perhaps. (3)

Benjamin Holt was similiarly fascinated by the early automobiles. His 1903 Oldsmobile was the first in Stockton, and in 1905 he built a small three-wheel automobile for running errands around the factory. (4)

Realizing the tremendous potential of gasoline motors, Benjamin and Pliny organized the Aurora Engine Company on October 6, 1906, with Pliny as president, Benjamin as vice-president, and Dan Gilmore as secretary-treasurer. Gilmore started with the Holt company as an office boy, later managed a branch office in Walla Walla, Washington, and returned to Stockton in 1903 to work as a salesman.

The decision to establish a separate engine company was based on the fact that some of the stockholders in the Holt Manufacturing Company were opposed to investing money in what was considered a novel and speculative venture. Accordingly, the forward-looking Holt officials organized a new company that could do experimental work without stockholder interference. The new engine plant, however, remained closely integrated with the parent factory, with its main office located at 325 Aurora Street.

Throughout 1906 and 1907, Holt and Aurora engineers, including Benjamin Holt, focused their attentions on building a gasoline engine track-type tractor. Field tests were conducted regularly as improvements were made, and as early as December 4, 1906, an entry in Company Secretary Dickenson's diary noted, "Tried the new Gasoline Traction Engine today. She is a grand success. Walks right along with harvester." (5)

With typical Holt business acumen, the Aurora Engine Company was organized just in time to ride the crest of the gasoline tractor tidal wave. Production figures suggest the Holts' rapid investment in the new line of agricultural machinery:

	Gasoline Tractors	Gasoline Motors
1908	**4**	**2**
1909	**53**	**59**
1910	**105**	**129**
1911	**298**	**290**
1912	**306**	**446**
1913	**262**	**366 (6)**

The Heer Engine Company of Portsmouth, Ohio manufactured this four-wheel drive tractor which had a 2-cylinder, 18-36 horsepower motor. The tractor had an enclosed cab and was designed to be steered by all four wheels. The solid-rim drive wheels could be replaced by open-cleat rims for work in soft rice fields. The tractor was awarded a blue ribbon for performance at the South Carolina State Fair in 1912. Some tractors were sold in Montana, Idaho and the Canadian Provinces.

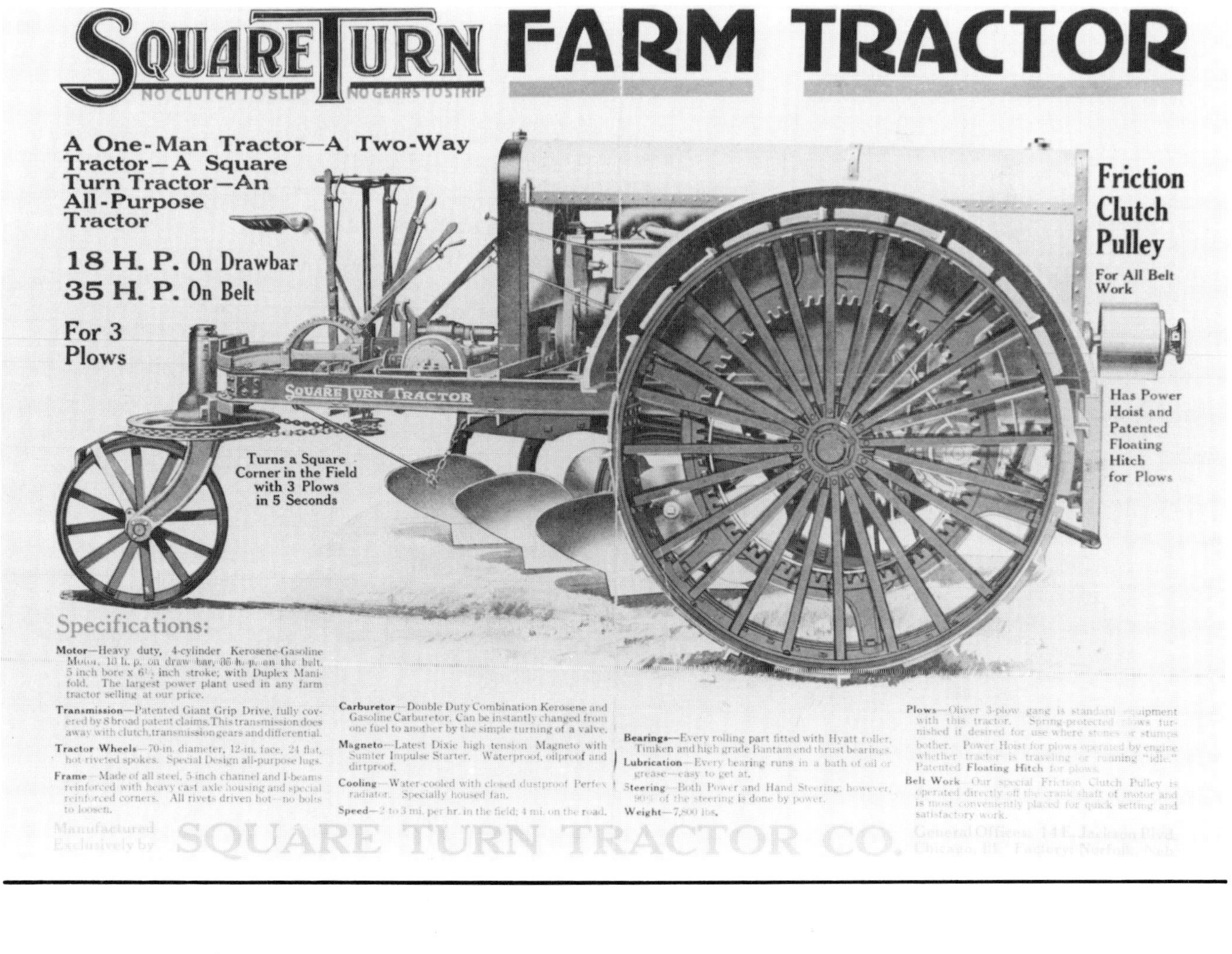

The Square Turn Tractor built in 1918 carried Oliver three bottom plows under the tractor frame. The motor which burned kerosene and used a friction clutch was equipped with a Dixie magneto, a Perfex Modine radiator and was powered by a Climax motor. Advertising stressed its ability to turn a square corner. An odd feature was that the driver could drive the tractor either forward or backward. To reverse, the driver simply reversed his seat and drove the tractor with the steering wheel in front instead of in the back.

SQUARE TURN FARM TRACTOR
NO CLUTCH TO SLIP — NO GEARS TO STRIP

A One-Man Tractor—A Two-Way Tractor—A Square Turn Tractor—An All-Purpose Tractor

18 H. P. On Drawbar
35 H. P. On Belt

For 3 Plows

SQUARE TURN TRACTOR

Turns a Square Corner in the Field with 3 Plows in 5 Seconds

Friction Clutch Pulley
For All Belt Work

Has Power Hoist and Patented Floating Hitch for Plows

Specifications:

Motor—Heavy duty, 4-cylinder Kerosene-Gasoline Motor, 18 h. p. on draw bar, 35 h. p. on the belt. 5 inch bore x 6½ inch stroke; with Duplex Manifold. The largest power plant used in any farm tractor selling at our price.

Transmission—Patented Giant Grip Drive, fully covered by 8 broad patent claims. This transmission does away with clutch, transmission gears and differential.

Tractor Wheels—70-in. diameter, 12-in. face, 24 flat, hot-riveted spokes. Special Design all-purpose lugs.

Frame—Made of all steel, 5-inch channel and I-beams reinforced with heavy cast axle housing and special reinforced corners. All rivets driven hot—no bolts to loosen.

Carburetor—Double Duty Combination Kerosene and Gasoline Carburetor. Can be instantly changed from one fuel to another by the simple turning of a valve.

Magneto—Latest Dixie high tension Magneto with Sumter Impulse Starter. Waterproof, oilproof and dirtproof.

Cooling—Water-cooled with closed dustproof Perfex radiator. Specially housed fan.

Speed—2 to 3 mi. per hr. in the field; 4 mi. on the road.

Bearings—Every rolling part fitted with Hyatt roller, Timken and high grade Bantam end thrust bearings.

Lubrication—Every bearing runs in a bath of oil or grease—easy to get at.

Steering—Both Power and Hand Steering; however, 90% of the steering is done by power.

Weight—7,800 lbs.

Plows—Oliver 3-plow gang is standard equipment with this tractor. Spring-protected plows furnished if desired for use where stones or stumps bother. Power Hoist for plows operated by engine whether tractor is traveling or running "idle." Patented **Floating Hitch** for plows.

Belt Work—Our special Friction Clutch Pulley is operated directly off the crank shaft of motor and is most conveniently placed for quick setting and satisfactory work.

Manufactured Exclusively by SQUARE TURN TRACTOR CO. General Offices: 14 E. Jackson Blvd., Chicago, Ill. Factory: Norfolk, Neb.

Within two years after the Holt Caterpillar crawler tractors were put in the market, the Holt Company had sold more of them than all their wheel steam traction engines produced in the previous 15 years. Of the first four tractors produced in 1908, three went to the City of Los Angeles for the Los Angeles aqueduct project, while the fourth, and the first used for farm purposes, went to Charles Lamb of Galt, California. (7)

While the transition from steam to gasoline power occurred quickly, it was a revolution that was several centuries in the making. Engineers, in fact, had worked almost 200 years to develop an adequate substitute for a steam-driven piston in an engine cylinder. Thomas Newcomen had invented a successful steam engine in 1712, yet it was not until the 1890s that pistons could be successfully powered by exploding gasoline vapor.

The first internal combustion engines used for commercial purposes were built by Jean Lenoir in Paris. In 1860, he patented an engine equipped with a carburetor which mixed vapor to explode in a cylinder when ignited by an electric spark. These engines burned the same gas used in gas lights, and because they did not involve compression of liquid gasoline to form an explosive vapor, they were practically noiseless. (8)

In 1867 Nicolaus August Otto, a shop clerk in Germany, designed an engine which received a gold medal at the Paris Exposition. Nine years later, he patented a relatively quiet four-stroke cycle engine which became the basic design used throughout the internal combustion engine industry. For this pioneering work he has been called the father of the automobile, tractor, truck and airplane. (9) The discovery of petroleum in Pennsylvania in 1859 and the development of electrical batteries, generators, and magnetos or small generators made possible the building of successful gasoline engines in the 1890s.

An example of the "Endless Track-Type" principle. The weight of the engine rested on steel rollers which ran on the inside of the tracks much like a locomotive on a railroad track.

First Holt crawler tractor used for farm purposes, December 14, 1908.

This Holt track-type tractor was built in 1911. The stationary attachment is the pulley used for belt work.

A modified Holt 60 horsepower tractor hauling baled hay about 1914.

Around 1915 a Stockton factory mechanic discovered he could move a tractor without the front wheel. To illustrate how well a track type tractor could function without this extra wheel the mechanic moved the tractor out of the building in a successful demonstration.

In the United States, the automobile and farm tractor appeared simultaneously. Charles Duryea drove his first automobile in Massachusetts in 1892 and in the same year John Froehlich built the first successful gasoline farm tractor in Iowa. (10) Following a period of trial and error, the Hart-Parr tractors produced in Charles City, Iowa, became the first commercially manufactured tractors. By 1903, fifteen of these were oil-cooled two-cylinder tractors which developed 45 horsepower and were in use on farms in the Midwest. In 1907, W. H. Williams, a Hart-Parr official, adopted the word "tractor" for these gasoline engines, to differentiate them from the earlier steam traction engines, and his simplified term is now part of the English language. (11)

When the Holts began marketing their first gasoline tractors in 1908, there were only 600 tractors in the United States, one-third of them Hart-Parrs and the rest divided among 30 different companies. Most of them imitated the steam engines already in use, being leviathan engines weighing from 20,000 to 50,000 pounds. Because many required factory-trained experts to start the motors, some farmers let them run overnight rather than face this baffling problem each morning. The Hart-Parr tractor instructional manual listed 19 steps for starting the engine and 13 for stopping it. Similarly, the Holt Manufacturing Company had more experts on the road starting farmers' tractors in 1910 than they had traveling salesmen. (12)

In addition, the repair bills for these huge machines were inordinately high, and the tractor companies received countless complaint letters. In 1908, for instance, a tractor owner in Illinois had run his engine only 30 days and complained that his repair bills were so great that he was ashamed to face his neighbors. Some farmers spent $1,500 a year on repairs, while bankers frequently refused credit to tractor owners because they were considered unstable and poor risks. The unreliability of these mammoth tractors coupled with the unwise overextension of credit by the manufacturers, led to the financial collapse of the industry in 1912, with many corporations going into receivership.

When measured against these events, the ability of the Holt Manufacturing Company to produce and market large gasoline tractors takes on added significance. Although company officials faced enormous engineering and financial problems, they succeeded where others failed.

Much of the popularity of the early Holt tractors can be credited to the engineering skill of Benjamin Holt and his associates. They avoided the bizarre tractor designs which produced ugly ducklings for many competitors, such as models with four-wheel drives, one big drum instead of drive wheels, and reins held by the operator who rode the plow instead of the tractor. (13)

Holt engineers also rejected motors bolted crossways on the frame, instead adopting the linear or automotive arrangement, which later became conventional on almost all tractors. The Holt tractor motors had overhead valves, a forced gear pump oiling system, and a chain drive rather than the more conventional three-gear reduction in the transmission. Some 18 feet high and weighing 18,600 pounds, the early tractors had a single front wheel to facilitate easy steering. (14)

On the new tractors, the substitution of tracks for the conventional wheels created complex engineering problems, which required immense experimentation and produced a voluminous patent literature. Pliny E. Holt designed and tried out several types of shoes for the crawler tracks, some with side holes as outlets for mud and some that wiped themselves clean as they revolved. Engineers calculated that the tracks usually exerted from 4 to 8 pounds of pressure per square inch. Company advertising brochures claimed that matching the bearing surface of a 24-inch crawler track would require a wheel 24 inches wide and 120 feet in diameter, and weighing 30,000 pounds. (15)

The design of the tractor's track mechanism also created a series of engineering headaches. When the first Holt track-type tractor was tested near the Stockton factory, it nearly shook itself to pieces. A reporter in "Sunset Magazine" described the incident:

> In this first caterpillar, the wheels which revolved the endless chain of tracks were of the same size, and the heavy tracks caused much of the shaking. Holt sent the tractor back to the machine shops, and in a short time it emerged for its second trial with one wheel smaller than the other. It moved without racking and shaking, pulled heavy loads, climbed steep grades and ambled awkwardly but without a pause over muddy and freshly plowed fields. "I've got it. I've got it" yelled Holt jumping up and down like an excited school boy. And he did have it. (16)

Because there were so many new details to work out on the gasoline tractors, daily affairs at the Aurora Engine Company were frequently hectic. An assistant engineer noted errors in original sketches of the tractor and urged that changes be made in engine bearings, crankshafts, and exhaust valves. Uncle Benjamin, according to his newphew, seldom had his promised new developments ready on time. Similarly, in the fall of 1910, engine production had to be curtailed because of the lack of connecting rods. Meanwhile, the new tractors at work in the Mojave Desert project needed an improved gear shift and steel pinions attached to the drive shafts. Connecting rods broke frequently, and magneto gears stripped their teeth. (17)

On occasion, the company made costly engineering mistakes. In 1909, when the Aurora Engine Company had been handicapped by an excess of broken crankshafts, Pliny Holt advised changing the forging process. Factory Superintendent Springer opposed this change because it would require the purchase of new dies

The "Half Breed" Holt tractor appeared in 1909. It was equipped with driving tracks on one side and a wheel on the opposite side. When attached to a road grader these machines were the forerunners of the motorized road graders.

Benjamin Holt designed this self-propelled combined harvester in 1911. Prior to 1911 gasoline motors provided auxilliary power for the Holt harvesters. Photo about 1916.

costing $700. Two years later, the crankshaft problem had grown much worse, and the company belatedly switched to the drop-forged types as Pliny had recommended. (18)

Occasionally, personnel problems affected workings at the Aurora Engine Company. When Benjamin Holt appointed Charles A. Tarbox to be superintendent of the Aurora, the new appointee immediately fired the popular head of the accounting department. Other employees lobbied on the accountant's behalf, but these pleas fell on deaf ears. Writing a hasty note to Pliny Holt, the accountant complained, "Old Tarbox is going to turn me out tonight, but I demand time to close my books. A man who is straight and who adheres to a straight line in his dealings is 'non persona grata' at this institution." (19)

Sometimes the Aurora Company suffered when talented individuals left to secure work elsewhere. One engineer resigned his engineering job in April of 1909 just when the firm was beginning to turn out engines in quantity. Pliny Holt offered to lighten his work load and increase his salary, but the engineer determinedly moved on to a position with less job pressure.

Despite these difficulties, the Aurora Engine Company's production was impressive. In 1906, gasoline motors were attached to several combined harvesters for use as auxiliary engines to drive the threshing mechanism, while in 1907 Benjamin Holt designed a small 12-25 horsepower crawler tractor for use on small farms. One of the inventor's dreams was to market a low-cost tractor comparable to Henry Ford's Model T automobile. The inventors Ford and Holt had met at a plowing contest in Winnipeg in 1910 and again at Panama Pacific Exposition in San Francisco in 1915. (20)

An experimental tractor was built at the Aurora factory in 1909, and was equipped with only one set of tracks instead of two. Hoping this "half breed" tractor would be suitable for work in orchards, Benjamin Holt was at first elated over its prospects and claimed it was a far better machine than anything he had designed before. Later the design proved to be a false lead. In 1910, another small Caterpillar crawler of 15-30 horsepower was manufactured, with over 300 of them sold for farm use. This suggests that neither Benjamin nor Pliny Holt were afraid to test innovative engineering ideas. (21)

Labor-management relations at the Stockton factory were generally good, except for crippling strikes in 1904. In February of that year, blacksmiths asked for the right to join the Machinist Union and establish an all-union shop. Holt executives refused to grant a closed shop and asked workers to pledge that they would neither join a union nor strike or boycott for a period of one year. The position of the Holt company was supported by Stockton business leaders and by the National Association of Agricultural Implements and Vehicles Manufacturers, which opposed the closed shop, the eight-hour day, and anti-injunction bills. (22)

When the blacksmiths refused to accept these terms, a general strike shut down the Holt Manufacturing Company on March 8, 1904. On the following day, fist fights occurred between the strikers and scab workers resulting in some injuries. Meanwhile, Stockton longshoremen, who sympathized with the strikers, refused to handle grain that had been harvested by machinery built by the Holt company. In response, Stockton warehouse officials locked out the longshoremen, and police were called in to prevent a riot. Tempers flared, fiery speeches were delivered, and violence was predicted. (23) The Pliny Holt family secured police protection for their children, Pliny G. and Frank Harrison, as they went to and from school.

After six months of turmoil, the Holt company workers lost the strike and the right to form labor unions. Secretary Dickenson noted in his diary that since the strike, the works had been run strictly on a non-union basis and "everything has proven that this is the only way to run our plant." (24)

For his part, Benjamin Holt concurred in this view. His attitude is revealed in a family story about a fracas which occurred during the heat of a strike. During a confrontation between strikers and non-strikers outside the Holt factory, a bitter argument ensued between two men. The non-striker took his horse and buggy to leave town, and his enraged opponent followed in pursuit. The non-striker than suddenly reined and hollered "whoa!". As his carriage came to a dead stop, the striker's horse and carriage crashed into it with such force that the striker catapulted into the air, hit the road, and rolled into a ditch. When Benjamin Holt heard about this episode, he roared with laughter so long that tears flowed down his face.

Apart from the months of labor dispute, occasional crises caused by engineering miscalculations, and the usual difficulties of setting up business in an infant field, the Aurora Engine Company enjoyed substantial profits. In October 1910, the company registered a profit of $32,000; two years later the assets of the company were $242,677 with liabilities of only $27,342. Since the firm had been organized only six years earlier with a capitalization of a mere $50,000, this was a fine financial performance. (26)

A gasoline-assisted combined track-type harvester built by Holt around 1919.

After the Holt Manufacturing Company purchased the Best Manfacturing Company of San Leandro in 1908 for approximately $1 million, thereby swallowing their strongest West Coast competitor, Aurora's machine shop was enlarged and new machine tools installed to meet the growing demand for tractors. Harold J. Baker, a machinist in the Aurora Engine factory, recalled that the Aurora Engine plant was a simple wood-frame building sheathed with galvanized sheet iron. During the winter months, heat was provided by hot air driven by fans. All machine tools were driven by flat belts powered by line shafts attached to journals (parts of the shaft supported by bearings) fixed to overhead beams. Most of the lathes, manufactured by Lodge and Shipley, had the ability to machine parts to a tolerance of .0001 inch using the high-quality jigs available at the time. Forty to 50 men worked in the shop. Common laborers received $1 a day for nine hours' work; machinists earned $3 and foremen $6. (27)

Machinist Baker attended regular company picnics in Oak Park, where employees enjoyed good food and athletic events, including a boxing match. In an age of business paternalism, factory hands often received a Christmas cash bonus and an occasional new silver dollar. Baker considered Benjamin Holt an intelligent man with sound engineering judgment, who, to his credit, never left the city where he made his fortune.

Another veteran Holt employee, Walter Gilgert, plowed in San Joaquin County with a gasoline tractor built by the Aurora Engine Company in 1909. He assisted another Holt employee who worked in the experimental division, doing custom plowing as an advertisement for the new tractors. Farmers were encouraged to stop and watch the newest model at work. As a plow tender on this outfit, Gilgert had the opportunity to see how the tractors functioned first-hand. Loose engine bearings and burned valves occasionally caused trouble, and at times the machines were hard to turn around at the end of a field because the front wheel slid forward rather than making the turn. Starting the motor was no problem—he simply thrust a starting bar into one of the holes in the rim of the flywheel and used this leverage as a crank. When the engine kicked over, the rod had to be jerked out quickly or the centrifugal force would throw it out at high velocity and pierce anything in its way.

As plowman, Gilgert was kept busy adjusting the depth of the plowshares, kicking out the stubble and tule plants which clogged the gangs of plowshares, or scraping mud which failed to scour from moldboards. On machines with headlights, he sometimes worked 24-hour shifts.

Occasionally working in the San Joaquin Delta as a plowman, Gilgert recalled that fires often burned underground, leaving dangerous pits which were invisible above ground. In these spots, the new crawler tracks bridged holes better than ordinary tractor wheels, which often became mired.

During the winter months, Gilgert worked at the Holt factory in Stockton, where he received $2 a day for 10 hours of work. Like most employees, he respected Benjamin Holt, in part because Holt often hired men who were down on their luck. If they were incompetent or did not work at their best for the company, Holt unhesitatingly let them go. First and foremost an inventive engineer and secondly a dedicated businessman, Holt was "a fine fellow to work for as long as you did what was right."(28)

Chapter 9

TRACTORS IN THE DESERT

It was Holt's Engine No. 1003 which first attracted national attention. This machine was ordered by the City of Los Angeles in 1908 for work on the mammoth Los Angeles Aqueduct Project.(1) Los Angeles voters had approved a $23,000,000 bond issue to finance the building of a 233-mile aqueduct across the Mojave Desert to the Owens River located on the east side of the Sierra Nevada. Under the direction of City Engineer William E. Mulholland, plans were made for building 30 miles of concrete-lined tunnels and steel syphons and a canal across the desert—an engineering feat of great magnitude, comparable, some said, to the building of the Panama Canal.

The project required the excavation of a long canal by using steam shovels. It also involved the transportation of huge amounts of supplies from rail heads across the desert and up the mountains over rough roads with grades as steep as 30 per cent. Faced with the task of moving enormous quantities of dirt and supplies, Mulholland naturally sought ways to reduce the cost of drayage, which for horse-teams normally ran from 30 to 40 cents per ton mile. When he heard of the new Holt "Paddle Wheel" steam engines, he conferred with Pliny Holt and placed an order for one on July 9, 1908.

An early experiment in hill climbing with a Holt track-type tractor took place in Stockton, California around 1906.

By early September, Engine No. 1003 was hauling materials from the railroad station at Mojave to the aqueduct, a distance varying from 4 to 12 miles. Mulholland's steam engine weighed 14 1/2 tons, developed 100 horsepower, and moved on tracks 90 inches in length. Since this steam-powered machine burned crude oil from a tank, only one engineer, rather than a full crew, was required to operate it, although boiler water had to be hauled in from great distances.(2)

Mulholland's initial response to No. 1003 were highly favorable, and he announced late in September 1908 that the solution to desert transportation seemed to have been found in the Caterpillar crawler tractor, which operated well on soft, sandy soils and on grades which were too steep for horses and old-style engines. The "road" to Sugar Loaf Camp in the Jawbone section of the aqueduct rose 1,000 feet in a six-mile stretch, a grade of 14 percent, he observed, but the steam engine hauled loads of 31 tons over this route daily with ease. At best, horses could pull only 600 pounds per animal and at a slower pace than the marvelous engine. He concluded, "It is expected that this engine will do proportionately well on level stretches hauling cement, sand, and rock. It is representing about 10 cents a ton per mile cost as against 30 cents for teams.

"(3) Because it made little sense to haul quantities of water to a steam engine in the Mojave Desert, and because Mulholland believed that mechanical power would be necessary to complete this monumental project, he sanguinely authorized the purchase on September 16 of a recently developed Holt gasoline tractor of the paddle wheel type. Two months later he ordered three more of these new heavy-duty gasoline track-type tractors. On November 1, Holt Company Secretary George L. Dickenson recorded in his diary: "Los Angeles City Council bought three more Caterpillar engines for $3,500 each, with Trucks and Trailers $600 each. The sales in Los Angeles the past week brought the local company $21,000."(4)

The new orders pleased Holt Company Officials. One wrote to a mining official in Nevada on February 19, 1909, stating that the company was selling the new gasoline track-type tractors faster then they could make them. He added, "We would be pleased to have you make a trip to Mojave, where we have three of these engines in operation and where we expect to send 24 more within the next few months."(5)

Just a month later, the official's predictions proved correct, marking the most significant contract to date for the Holt Manufacturing Company. The Los Angeles Board of Engineers and Board of Public Works had called for bids to supply tractors and steel dump wagons for work on the aqueduct. Although the Best Manufacturing Company of San Leandro had underbid the Holt Company on the contract for dump wagons, Holt assistant factory manager R.S. Springer convinced the Los Angeles authorities that the Best wagons would violate patents held by the Holt Company, a claim of dubious validity. After a tense three-hour session on March 23, 1902, the contracts for the tractors and wagons were awarded to the Stockton corporation. Springer's jubilant telegram home read: "Largest single contract which will be awarded in construction Los Angeles Aqueduct was awarded today to the H. M. Co., Stockton, for 25 Caterpillar traction engines and 80 wagons, amounting to over $141,000."(6)

Work on this new contract rushed along without delay. By early June, all departments in the Stockton factory were operating from 6:00 a.m. until 6:00 p.m., while a night shift running from 5:00 p.m. until 3:00 a.m. had been added in the tractor department.(7)

Holt steam track-type tractor hauling freight on the Los Angeles aqueduct project in 1908.

In southern California, many of the Holt employees supervising the operation of the Holt tractors stayed at the Harvey House at Mojave where the Santa Fe and Southern Pacific Railroads maintained freight yards. Normally Mojave housed only a handful of people, but during the aqueduct construction the town boomed until it supported 15 saloons. A place known as "Smitty's" gained special attention because whenever greenhorns offered to buy drinks on the house, Smitty would step outside and ring a large bell that brought most of the citizenry rushing through the tavern's swinging doors.

By all accounts, Mojave was a tough town, tougher than most other frontier mining, oil, and railroad towns including Goldfield, Ryolite, and Tombstone. One bartender recalled he had to fight with fist, bottles, and gun to hold down his side of the bar whenever "that crazy mob of hop heads, drunks, mule skinners, and toughs came at me right over the bar without any warning." He also claimed to have picked up "a dead man or two every morning when daylight revealed bodies lying in streets, alleys, or vacant lots." One worker described a three-month period when an average of eight men were killed each week.(8)

Although Holt officials were delighted to sell 28 tractors for use on the desert aqueduct, a multitude of headaches for the company followed the sales. The first steam engine that went to work on the Jawbone Division hauling freight to Pinto Hills, Midway Canyon, and Sugar Loaf cost 67 1/2 cents per ton mile in its first month of operation, double what horses cost to maintain. As a result, this machine was abandoned.(9)

The gasoline track-type tractors were an improvement on the steam track layers because no constant supply of water was needed. The gasoline machines soon evidenced serious problems in this difficult proving ground. They encountered intense heat in the valleys, rough terrain in the foothills of the Tehachapi Mountains, and winter snows at high elevations. As the tractors churned over rough roads, the whirling sand created excessive wear on moving parts and necessitated an inordinate number of repairs. Cast-iron gears had to be replacated in steel. Furthermore, the engines' original suspension systems proved to have inadequate springs, causing severe engine damage, and the machines' two-speed transmissions did not have an extra-low gear needed for the heaviest work.

Paul E. Weston, the Holt Company service expert sent to Mojave, complained to the home office several times about the problems of these tractors and demanded faster service in furnishing repair parts. From his

Holt tractor at work on the "jawbone division" of the Los Angeles aqueduct project in 1909. This picture appeared on the front of "Scientific American" on March 6, 1909.

perspective, the executives in Stockton were insensitive to the pressing needs in the field.

In a defensive mood, factory assistant R.S. Springer replied to Weston on February 13, 1909, that it was impossible to make changes in the tractors in less than six weeks time. It had been difficult enough to make deliveries on time because they had been forced to redesign most of the engine for the demands of aqueduct work. The change to a three-speed transmission, for example, had required new sketches from the drafting department for new patterns and castings. These changes necessitated others in the size of the main drive pinions, which in turn required alterations in the design of the frame. He added, "You have been with the company long enough to appreciate how much work is required to design an entirely new engine, and we trust that you will be able to make this matter clear to Mr. Mulholland."(10)

Harassed by the Los Angeles engineers for failing to meet production schedules, Springer sent a memorandum to the company's tractor division on April 2, 1909, about the six all-steel tractors scheduled for completion in early April and the additional tractors promised in August. Obviously anxious about meeting the deadlines, Springer admonished, "We have furnished a bond of approximately $30,000 guaranteeing that we will make delivery on specific dates, and we certainly will expect you to do your part and deliver these engines completed and tested on these dates."(11)

Summarizing the Holt factory's production problems in a letter to Pliny Holt on April 23, 1909, Springer reported that Benjamin Holt had just returned from an inspection trip to Mojave to report that all was not going well. Holt had noted that the new three-speed transmissions were wearing out because some drivers ran the tractors in low gear all the time rather than only in emergencies. While the Los Angeles engineers thought the friction clutches might drag and wear out the camel-hair lagging in a short time, their fears were exaggerated. In addition, Springer reported cooling fans had broken because they were not well balanced, and after a flywheel had broken on Engine No. 1002, Benjamin wanted the hubs on the flywheels made stronger. An auxilary fuel tank also was proving to be imperative, because when engines made steep climbs, fuel would not reach the carburetor. All the machines' frames needed to be reinforced, braced, and made stronger in several points.

Holt had also noticed, Springer continued, the engine tracks showed excessive wear, suggesting either that the pins lacked adequate lubrication or that they were made from inferior metal. In addition the track shoes bent too easily and needed to be case-hardened. In concluding his

Holt tractors were used to haul freight across the Mojave Desert around 1909.

report Springer confided to Pliny Holt that in many cases poor materials had been used on the aqueduct machines. "You may remember that the shoes on Engine No. 1007 were made of scrap tank steel, most of the material being second-hand stuff which had gone through the San Francisco fire of 1906 as part of the jail cells in the Hall of Justice. This material, of course, was naturally soft, and this may be the reason why they wore out so fast."(12)

When the performance problems of the engines continued into the late summer months, Benjamin Holt, his nephew C. Parker, and Springer personally visited the aqueduct work to see the operations first hand. Despite the personal inspection of the machines and operating conditions and their recommendations on returning to Stockton, grave engineering problems persisted. In the fall of 1909, Springer wrote to J.B. Lippincott, an assistant to William Mulholland, to acknowledge receipt of the aqueduct cost report on engines working on the Jawbone Division. The report showed that the excessive cost of 56 cents per ton mile had been due primarily to the high cost of repairs on the machines. In conclusion, Springer plaintively promised to ship more spare parts, to set up additional repair shops at the work sites, to exchange tractors needing to be rebuilt, and to send three more mechanics to Mojave to assist with maintenance work.(13)

Although the Los Angeles Aqueduct Project—finally completed in 1913 after monumental construction efforts—was ultimately profitable for the Holt Manufacturing Company, the final project engineer's report published by the city in 1916 assessed that the 28 Caterpillar crawlers used on this project had been failures. The report stated that transportation had been the most difficult aspect of the project because of the enormous volume of freight to be hauled over long distances and the scarcity of water in the region. As work had progressed on the project, two factors had become apparent: first, as the tractors aged, the number of breakdowns greatly increased, running up the cost of repairs and increasing significantly the freighting costs; and second, the cost of freighting by horses had been reduced so as to make horses more efficient than machines. Teams as large as 14 animals could be handled by one driver on desert roads, making it possible to cut the cost of team-hauling to 12 or 13 cents per ton mile, one-half the cost of tractor-hauling. While new engines could operate at the same low cost, the cost of their repairs made mechanized freighting twice as expensive. As a result, the official city report concluded:

> After an earnest effort to make a success of these engines, they were finally abandoned, and the gas engines taken from there for uses in places where gasoline power was desirable. The frame work of some of these engines was used for steel forms for concrete work, and some "Caterpillars" were sold to private parties for farming purposes. They were the only type of equipment that was purchased in the building of the aqueduct that was unsatisfactory.(14)

ORE WAGON

Design	4-wheel, 8-ton, side dumping
Capacity of Box	4 cubic yards
Weight (approximate)	6450 pounds
Front Wheels	diameter 4 feet 0 inches; width 12 inches
Rear Wheels	diameter 5 feet 4 inches; width 12 inches
Width of Track (over all)	6 feet 2 inches
Distance between Axles	11 feet 3 inches
Radius of Turning	27 feet

A Holt Manufacturing Company ore wagon. The four-wheel 8-ton wagons could be tilted to facilitate dumping of the dirt, gravel or rock. These were used on the Los Angeles aqueduct project in the Mojove Desert.

On the positive side for the Holt company, according to Holt employee Paul E. Weston, the first steam engines used on the Mojave Desert performed better than any others in previous history. The gasoline-engine tractors also did well in the desert dust, considering that air cleaners and oil filters had not yet been invented. Finally, Weston remembers:

> We pulled some un-godly loads over those mountains which could not have been handled in any other manner. We took those big 50-ton Marion steam shovels and Ingersoll-Rand compressors up those steep grades and across deserts without their having to be torn down and re-assembled, as would have been done without the new Caterpillar power.(15)

While the Holt company's record was far from one of unqualified success in the years 1908 through 1912, the nation's tractor industry suffered an even worse fate and completely collapsed. Most manufacturers of large farm tractors either filed bankruptcy or converted their plants to build lightweight tractors. Famous tractor companies such as Kinnard Haines, Charter Engine, Aultman-Taylor, Emerson Brantingham, and many others disappeared because the high cost of repair on their big

Holt track-type tractor pulling a Holt ore wagon.

tractors made the machines uneconomical to produce and market. When 248 tractor owners were querried about the nature of their machines' repair needs, one Montana farmer tersely answered, "Everything." Bankers compounded the problems for tractor companies by refusing to loan money to farmers who had sunk several thousand dollars into the new-fangled tractors. Eventually, most of these big early tractors were discarded, with as many as 500 of them dumped into a single graveyard. In fact, it took years for companies to sell off their inventory of the 12,000 large-size tractors produced in 1912.(16)

Meanwhile, the Holt Manufacturing Company survived the tractor crisis of 1912 and became one of the few corporations to weather the hazards of the early experimental work in producing these machines. In the first place the Holt company's contracts with the City of Los Angeles for work on the aqueduct brought in $141,000 in cash which helped the firm during these crucial times. In spite of the mediocre performance of crawler tractors on the aqueduct product, the Holt company received much favorable publicity. One company official estimated that the announcement of the sale of 28 Holt tractors to the City of Los Angeles in 1908 was worth $20,000 in advertising.(17) One example of the favorable press appeared in the "Stockton Daily Evening Record" on September 25, 1908 when the editor predicted that "the Caterpillar will without question become the greatest advertisment that Stockton has ever had for achievement in mechanics, not excepting the combined harvester, for the new traction engines will be used in all parts of the world, in the tropics and in the frozen North, and every one of them will bear the label of Stockton manufacture."(18)

Furthermore, the Los Angeles Aqueduct Project had provided a good proving ground for the Holt tractors. The contracts included deadlines which forced Holt engineers to find solutions to mechanical problems quickly and out of the Mojave Desert experience came technological improvements such as all-steel construction, better spring suspension systems, better clutches, three-speed transmissions, and above all, durable, strengthened tractor parts. The final result was a successful track-type tractor that could be used on American farms, as well as for road building, lumbering, mining and freighting.

Chapter 10

BRINGING THE GASOLINE ENGINE TO RURAL AMERICA

Although central to the economic life of Stockton, the Holt Manufacturing Company and the Aurora Engine Company were only part of a nationwide tractor industry which sprang to life around the turn of the century. As is customary in any newly expanding industry, fierce competition for markets developed almost overnight among several hundred eager manufacturing companies. In the early years, success meant substantial profits and failure meant bankruptcy, and under this pressure, the relatively new pursuit of mass advertising became extremely important. In the tractor trade, advertising included circulars, catalogues, calendars, letterheads, lithographs, repair lists, engine engravings, and letters. Other unusual advertising gimmicks distributed to would-be customers included watch fobs and stick pins designed in the shape of a particular model tractor. Some salesmen carried a threshing cylinder tooth ground to a razor-like edge which, they claimed, showed the fine quality of the steel used in its manufacture. Many ads appeared in new magazines published especially for readers interested in power farming, such as "The American Thresherman" and "The Gas Review".

For the Holt companies, it soon became apparent that the Caterpillar trade name was especially valuable as an advertising and public recognition device. Accordingly, the company registered the word as a trademark in European countries, Africa, and the East Indies. Translating the term "Caterpillar" into foreign languages resulted in "Rups" in Dutch, "Larve" in Norwegian, "Larv" in Swedish, "Toukka" in Finnish, "Oruga" in Spanish, and "Lirfa" in Icelandic. (1)

To attract wide attention, the Holt salespeople regularly entered their tractors in popular pulling contests where farmers who already owned machines could cheer for their respective sides. Company servicemen backed their machines with bets and used every trick to try to gain an advantage. One Best Company employee who took part in such a contest in 1914, secretly bored the cylinders of the 75-horsepower Best engine to make them one-eighth inch larger. He recalled:

> I was driving the Best when I noticed the clutch began to slip and smoke. C. L. Best was right beside me in the heat of battle. We were pulling one of those big John Deere ten-bottom plows set deep. Out of the corner of my mouth I said to C. L., "We've got to stop or we'll burn out the clutch." "Keep the s.o.b. going, never mind the clutch," he snapped. So we finished the trial strip, but the clutch was gone. We skinned the rump off the Holt that time. (2)

During the horse-automobile controversy the pro-horse literature often carried cartoons to lampoon the pro-auto crowd. Cartoon from "American Horse Breeder," November 17, 1903.

Always looking to the future for his company, Benjamin Holt became a strong advocate of advertising. He exhibited his tractors at state fairs, where they performed before large crowds. Some companies had young boys operate the machines to prove their ease of handling. Another demonstrated the smoothness of its tractor's clutch by opening the back of a watch, tying it to a post, and then backing the tractor against the watch to snap it shut without cracking the glass face of the timepiece.

The greatest publicity for new model tractors was generated by the summer plowing contests held near Winnipeg, Manitoba, from 1908 to 1913. Invitations were extended to scores of manufacturers of steam and gasoline engines. Railroads offered special passenger fares, while local businessmen provided entertainment for visitors. Because the prize winners received international publicity, the competition to win these laurels was perhaps the toughest in the annals of mechanical power in American agriculture. (3)

During these trials, steam and gas tractors surged across the prairies pulling plows toward flag markers a mile distant. As the engineers struck their furrows, agricultural college professors serving as judges recorded the performance of each machine in great detail. When this data was compiled, the winners were announced across the country by telegraph. These yearly contests brought important leaders of the farm power industry together, such as J.B. Bartholomew, president of the Avery Company, Henry Ford, and Benjamin Holt. The results of these major social and technological events demonstrated convincingly that the gasoline tractor was destined to replace steam engines on farms of the future.

The Holt Manufacturing Company received a number of important awards which increased its good reputation. In 1904, Benjamin Holt shipped a steam plowing outfit to Spain, where it won special honors in an exhibition near Seville. In a plowing demonstration, the King of Spain waved his arms to signal the start of a field test and then, much against the wishes of his prime minister, he followed the plow 200 yards down the field. (4)

Similarly, the Holt tractor was awarded cash prizes for excellence by the government of Argentina in 1911 and 1912, as well as first prize in a contest held in Tunisia in 1912. When the Russian government held an exhibition near St. Petersburg under the patronage of Emperor Nicholas II in 1913, five German and four American tractors were entered in the trials. All the first prize gold medals were awarded to the Holt Manufacturing Company. (5)

Another important exhibition opportunity occurred at the Panama-Pacific Exposition held in San Francisco in 1915 to celebrate the completion of the Panama Canal. At the Palace of Agriculture Hall, the Holt company displayed four tractors, as well as a combined harvester, plows, harrows, land levelers, and a large sage-brush clearing machine. Near the Marina, a 75-horsepower Caterpillar crawler climbed a steep incline to demonstrate its hill-climbing ability. The Holt Manufacturing Company received grand prizes for the crawler tractor and for the combined harvester and five Medals of Honor and two Gold Medals for other machines—the highest awards made in each category.(6)

In addition, Holt tractors were entered in local tractor shows held in such places as Fremont, Nebraska; Hutchinson, Kansas; Champaign, Illinois; Spokane, Washington; and Aberdeen, South Dakota. These demonstrations were sponsored by manufacturing firms who were eager to exhibit their tractors in field tests. These shows were prominent in various parts of the country from 1913 to 1919. Since rural people were interested in observing gasoline tractors as a new source of power, they flocked to these demonstrations in great numbers. During the tractor show in Fremont in August of 1916 daily crowds reached 60,000 for each day of the

The Acason Farm Tractor Company, Detroit, Michigan built this 1915 Model T which sold for $265. Automobiles were used for farm field work by adding attachments.

four-day event. The "Fremont Herald" on August 10, 1916 ran the banner line, "Enormous Crowds at Tractor Show" and stated that there were 8,000 automobiles parked in an 80-acre field. During this exhibition, 50 different tractor companies furnished 250 tractors used to plow 1,200 acres of land. The value of all machinery on the grounds was placed at one million dollars, while sales reached $1,300,000. Farmers tramped up and down the fields watching the tractors perform and many bought tractors on the spot. (7)

The arrival of gasoline tractors and automobiles created a serious debate between pro-horse and pro-machine factions. At first the owners of new horseless carriages who drove from the city onto country roads were resented by many rural Americans. The owners of these "Red Devils" were called reckless trespassers, and farm journals often featured headlines such as, "The Murderous Automobile" or "The Deadly Auto."(8)

Furthermore, farmers usually liked horses and many thought brainless machines were inferior to intelligent animals. Angered by auto traffic in the country, some stubborn farmers dug ditches across public roads or blocked them with logs or cables.

The auto proponents defended themselves against charges that they were maniacs who left people dying by roadsides. While autos, they granted, might frighten horses, so did railroad locomotives, street cars, Fourth of July celebrations, and lots of other things. With the use of automobiles, horses would no longer have to fall on icy streets or die from the heat on torrid summer days. Finally, the auto crowd pointed out that horses were expensive to own, requiring five acres of grain and grass for annual maintenance. "Why can't we use the ground for something we can eat ourselves?" complained one Texas farmer. (9)

A similar debate accompanied the arrival of the gasoline tractor. The horse-tractor controversy appeared in farm journals, in newspapers and at literary society meetings in one-room school houses. The key issues were economic: Would tractors do the job and would they be worth the expense?

Horse lovers insisted that horses were less costly than tractors, and they reproduced themselves. As yet, they quipped, no farmer had ever gone out in the yard and discovered that his tractor had given birth to a tractorette. Furthermore, since horses lived off the land, they did not require any outlay of cash for fuel. In addition, tractors were unreliable, they packed the soil, mired down in soft spots, and made loud roaring noise.

Some business leaders feared that the purchase of tractors would endanger the credit of the nation and create tight money at the banks. Tractors would tie up capital, pile up debts, and make money scarce in the marketplace. Tractors might be luxuries like polo ponies and private schools for girls. Besides, they snorted,

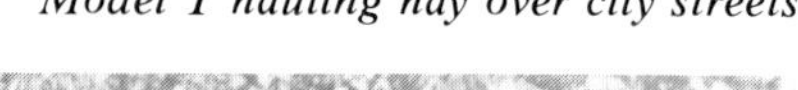

Model T hauling hay over city streets.

choked, and gurgled "like a hog's snout in a trough of buttermilk."(10) Some farmers were so tractor crazy that they borrowed from banks, sold livestock, or plastered their homes with mortgages.

Apologists for tractor owners proved equally vocal. They insisted that tractors were needed in farming, and that if Eastern financiers were worried about public extravagance, they should examine the brewery industry. A writer for "The Gas Review" in 1910 observed that bankers posed as apostles of the Almightly appointed to guard other people's money, but that this whole crowd had better put mufflers on their exhausts. (11) Some tractor enthusiasts taunted, "You can't fix a dead horse with a monkey wrench," and cited the horse plague which killed 30,000 horses in Kansas in one year. (12) Others called attention to horses with heaving flanks and scalded shoulders as they worked in 100 degree weather, or about burying those that dropped dead from sunstroke.

For their part, Holt officials stressed the advantages of tractors over horses. Tractor power reduced the drudgery in farm work and encouraged young men to stay on the farm rather than drift off to the city. Then, too, tractors would eliminate transient laborers, thus getting rid of "King Hobo" who worked when he pleased and got drunk when he was needed most. One Holt bulletin claimed that tractors gave farmers a chance to sign a Declaration of Independence from men who said, "I Won't Work, I Won't Wash, I Want Whiskey," a veiled reference to the Industrial Workers of the World, a radical labor group. (13) Furthermore, a tractor would plow, harvest, saw wood, grind feed, and shell corn without asking for a holiday, shorter hours, higher wages, or easier work. Above all, a tractor would not step on farmer's toes or switch a tail in his face.

After the outbreak of World War I in 1914, the editor of the "Caterpillar Times" pointed out that the European conflict had created a great demand for horses, with some contractors shipping as many as 46,000 head in one lot. An editorial admonished, "Now, brother farmer, think it over. It is not a question of tractors coming in. The tractors are here." After stating that tractors were a good investment, the writer listed 40 different tasks which could be handled by a Caterpillar track-type tractor. (14)

Convinced that power farming would replace horses, several manufacturers offered classes to train men to operate tractors. In 1914, the Holt Manufacturing Company established a Caterpillar Tractor School in Spokane, Washington, and in the following year one was held in Stockton. Instruction was provided by professors from agricultural colleges and experts from the company. Owners of Holt tractors and combined harvesters could waive the $25 tuition fee. Over 100 students enrolled in these courses in 1915.

By the 1920s, the winner of the tractor-horse controversy was obvious and the automobile, truck and tractor have gradually displaced the horse from the rural scene. Today, the United States Census Bureau no longer counts the number of horses and mules at work on the American farm because it is inconsequential. This part of the agricultural revolution is complete.

This 1919 photo shows experimental tracks attached to a Model T Ford.

Chapter 11

THE PEORIA STORY

Following the Holt Manufacturing Company's decision in 1906 to set up the Aurora Engine Company for developing and producing gasoline engines, the parent company resolutely committed its resources to perfecting the Caterpillar tractor and marketing it across the nation. This called for considerable expansion of its already sizeable plant facilities in Stockton and modernization of its corporate operations.

The Catholic cemetery block on the west side of Aurora Street had been purchased and added to the original factory site in 1898, and in 1902 a new foundry had doubled the company's capacity for making castings. In addition, just after the turn of the century, the first Holt building erected in 1883 was remodeled into an impressive office headquarters. A visitor to the plant in 1906 noted that telephones were used for conducting business within the plant, and dictaphones, which recorded the speaker's voice on a cylinder, were operated by company secretaries. With awe he added, "It is simply marvelous. System prevails everywhere." (1)

Following the destruction of manufacturing facilities in San Francisco in the 1906 Earthquake and Fire, the Holt companies in Stockton were left holding the most complete stock of iron and steel anywhere on the Pacific Coast. According to one Stockton reporter, Holt could supply its far-flung customers with everything from tacks to traction engines: "Their market is the world."(2)

In 1909, Holt executives took a hard look at the national economy and recent developments in agricultural mechanization and production, and tried to realistically assess the options of continued expansion versus maintainance of the present level of company operations. All signs pointed to a boom in the use of farm machinery. Farm acreage had doubled since 1880, with grain harvests reaching five billion bushels annually. The use of improved farm machinery after 1870 had increased the productive capacity of farm workers by 86 percent, and since 1900 a period of farm prosperity had generated a widespread spirit of optimism. (3)

Assessing the production aspect of farm machinery, the Holt executives knew that some 37 companies were manufacturing farm steam engines in the United States, and most of them were rapidly converting their production to gasoline tractors in the race to grab a share of the new market. The J. I. Case Company in Racine, Wisconsin, had doubled its sales of farm machinery in the last decade, while the Minneapolis Threshing Machine Company had reached sales of $3 million annually. The Rumely Company in Indiana was tooling up to manufacture 50 tractors a week, and the Hart-Parr Company in Iowa planned to issue 1,000 tractors each year. (4)

With their competitive spirit piqued by these developments, the Holts decided to expand operations by establishing a factory in the Midwest, a move announced by the "Stockton Independent" on January 19, 1909. Benjamin's nephew Pliny was to be in charge of the plant, and he chose Minneapolis as a base for this expansion venture, in part because he had lived in the city and received his engineering training at the University of Minnesota. In March the Northern Holt Company was incorporated with Pliny Holt as president and treasurer and his brother Ben C. Holt as vice-president. Two months later, a contract was signed with the Diamond Iron Works in Minneapolis to assemble Holt's first ten tractors from parts shipped from Stockton. (5)

Shortly after arriving in Minneapolis, Pliny Holt and his old friend Charles J. Gotshall made a trip to Winnipeg to explore farm machinery business operations in Canada. Pliny's interest in the Canadian market no doubt reflected the enthusiasm in the Twin Cities of Minneapolis and St. Paul for the agricultural potential of the Canadian provinces. A half million people had emigrated to Saskatchewan from the United States during the previous ten years, and real estate offices in

Minneapolis were selling 20,000 acres of Canadian land daily.

As a result of the Canadian trip, the Canadian Holt Company, Ltd. of Winnipeg was set up, with all Canadian rights to be owned by the Northern Holt Company. (6) The contract between Pliny Holt and Benjamin Holt in 1909 stipulated that Pliny must sell more than 51 tractors in Canada within the first three years. If he failed to do so, the right to make sales in Canada would revert to Benjamin. Pliny was to receive $300 commission on each tractor sold in Canada. (7)

Despite Pliny's reasoned optimism about expansion of the Holt market into the Midwest and Canada, he immediately ran into the nemesis that plagues all new manufacturing businesses—the need for cash. Great quantities of capital were required to buy, equip, and operate a factory for producing Caterpillar track-type tractors, and Pliny and his brother Ben needed lots of money quickly.

This cash shortage handicapped the Northern Holt Company in its first year. The 1909 ledger listed office expenses of $11,980, and income from the sale of two tractors at $7,016, with a loss of $2,963. (8) The Northern Holt Company immediately tried to sell stock so as to secure enough capital to operate the plant until sales could provide ongoing funding. But, as Pliny discovered, it proved almost impossible to sell stock in a company whose products were not known in the area. Selling bonds was also difficult because it had not been adequately demonstrated that a crawler tractor would perform successfully in Midwestern soil conditions, and would-be investors believed it impossible to sell Caterpillar tractors for $4,000 when similar Hart-Parr tractors were selling for $2,650.

Concerned about this crippling financial bind, Northern Holt Vice-President Ben C. Holt wrote Pliny on April 14, 1909, from his office in Walla Walla, Washington, that he had been unable to get any cash for the Northern Holt Company from the Stockton home office because there was nothing available to be let out in loans. Nor could he himself forward funds because he had already exhausted his credit with local banks. He urged Pliny to reduce salaries and cut expenses in Minneapolis, reflecting that he could not remember a time when his business affairs had been so tight:

> I do not see how I can finance a company so far away. I have heavy obligations to meet and no money to meet them with . . . C. Parker just informed me that you have overdrawn $5,000 since January. Don't say anything. I merely let you know that I am in this fix. (9)

At this crucial point, a farm implement dealer in Peoria named Murray M. Baker entered the picture, sparking a series of events which affected the entire future of Holt operations in the Midwest. Baker was ambitious, shrewd, enterprising, and, importantly for Pliny Holt, always eager for a new business venture. Baker's scheme involved a defunct farm machinery factory in Peoria. (10)

In 1897, industrialist William H. Colean had erected a modern factory in Peoria for manufacturing steam traction engines and threshers. When he failed to shift production from steam to gasoline engines, his company went into bankruptcy. The closing of this factory in 1908 was a serious blow to Peoria, and the factory's creditors were eager to sell the building and equipment to investors who would pay off the old debts and establish a new business in the city. Accordingly, Baker suggested to his friend John M. Olmstead that he stop in Stockton on his upcoming western vacation and see if the Holt organization would be interested in purchasing the defunct Colean Manufacturing Company. (11)

As suggested by Baker, Olmstead called on his college friend C. Parker Holt in San Francisco and described the Colean factory. C. Parker immediately wrote Pliny in Minneapolis about Olmstead's visit. In turn, Pliny wrote Baker on May 13, 1909, that he understood the Colean factory which had cost $500,000 to build, had gone "co-fluey" and could be bought for 20 cents on the dollar. Pliny concluded, "I am interested in a plant for building a new line of traction engines which we are preparing to put on the market in this territory. Any information you can give me will be appreciated."(12)

In response, Baker wrote several letters to Pliny Holt, commenting that the Colean plant would be a real bargain when sold at auction on July 13, 1909. Since several groups of investors were likely to make strong bids, he urged Pliny to visit Peoria to evaluate the property.

Following this advice, Pliny arrived in Peoria on June 22, 1909, where he accompanied Baker on a tour of the factory. Impressed with what he saw, he wrote an enthusiastic letter to Ben C. in Walla Walla. The three-story brick building was equipped with modern machine tools, and four railroad side-tracks connected with 13 major lines, while steamboats on the Illinois River provided additional transportation. Pliny estimated it would take $250,000 to build similar facilities and this opportunity was just what their company needed to get off the ground in the Midwest. (13)

In mid-July, Baker wrote Pliny that it would probably take $75,000 to purchase the factory and the 10 acres of land on which it was located. Pliny thought this price was acceptable, providing the plant could be purchased on favorable terms since it would be difficult to raise cash on short notice. As understood by the Holt officials, the Colean factory could be bought for $75,000, fundable by floating new bonds and requiring no cash.(14)

In Stockton, Benjamin Holt favored the move, because he believed a first-class factory in Peoria would be an ideal center for doing business in Illinois, Iowa, Indiana, and Missouri, potentially good states for Caterpillar crawlers. Ben C. Holt in Walla Walla concurred, convinced that "this production is the one to stick to." Also pleased with the negotiations, Pliny enthusiastically wrote to Baker on July 17:

This picture taken in the 1920s includes many of the most important associates of Benjamin Holt. Left to right: Parker Holt, Ben C. Holt, Russell Springer, Charles Neumiller, Murray M. Baker, Thomas Baxter, Pliny E. Holt and Dan Gilmore.

I am sure that this deal marks the beginning of one of the largest enterprises in the Middle West, and assures the city of Peoria of an industry that they will be proud of in the future. If this deal goes through as planned, I intend to build practically all of the parts for our machines for both the Middle West and Pacific Coast territories and this will soon call for a material enlargement of the present plant.(15)

Then negotiations hit a snag—not about the price, but about the terms of sale. When the Colean factory was sold at auction on July 15, 1909, it was purchased by the Dime Savings & Trust Company of Peoria for $52,600 with interest at 6 percent until date of redemption. This was a bargain, because experts had appraised the property at $168,000. As planned Baker then offered $10,000 to the Dime Savings & Trust Company for the sale certificate, but now the creditors asked for an additional down-payment of $20,000 in cash. After negotiations, the best terms Pliny Holt and Baker could secure from the bank required the payment of $30,000 in cash, with the balance of $45,000 to be paid within one year.(16)

Pliny Holt began to get cold feet, expressing his reservations to the Stockton office. He opposed pouring so much cash into this venture because they needed all their funds to operate the main factory and the branch in Minneapolis. He also knew that it would be impossible to secure loans from Peoria bankers unless the Holt Manufacturing Company put some money into the deal. Turning pessimistic, he said he preferred a factory in Moline or Rock Island, although the Colean plant was worth twice the asking price. In response, Ben C. Holt forwarded his advice in a telegram on August 2: "Wouldn't hurry. Advise rejecting temporarily." Pliny concurred, sending a letter the following day to the Dime Savings & Trust Company with the unsigned contract and a note that he did not need a factory in the immediate future and would be looking at other attractive prospects. The deal was off.(17)

Efforts then were made to finance a factory at some other Midwestern location. Chicago financier William Stone agreed to make an investment if he were made a majority stockholder, but Ben C. Holt refused the offer, saying that the Holts had invested $200,000 in perfecting the tractor and would not be relegated to being part-owner of a branch factory.(18)

While hoping that more favorable terms might be secured from the creditors of the Colean Manufacturing Company, the Holts received word from Baker on September 2, 1909, that the J. L. Owens Company of Minneapolis had offered $75,000 for the Colean factory and a payment of $11,500 for the purchase certificate. On the following day, a firm representing J. B. Bartholemew of Peoria offered $13,000 for the sale certificate. Under this pressure, Baker purchased the certificate for $11,500, forestalling the overtures of the other two bidders and buying additional time for the Northern Holt Company. (19)

Baker knew that the Bartholemew offer was no idle threat. Bartholemew was president of the Avery Manufacturing Company in Peoria, a firm which was turning out one farm steam engine every five hours and a threshing machine every two hours, with total annual sales of $10 million. By wire, Baker asked Pliny Holt if he would accept the terms of sale: $11,500 for the purchase certificate in 30 days, $44,000 on November 1, 1910, and $19,500 within two years—with a total cost to be $75,000. On September 29, 1909, Pliny telegraphed his reply: "All right if necessary, subject approval of our people. Don't like their attitude." (20)

Writing Stockton to explain his need to act without company approval, Pliny urged, "I feel sure we are making a good investment and one that is the beginning of a tremendously big industry, the profits of which will give us all big returns for the time and thought we have given toward getting the business started." He explained the final terms of the sale—and requested:

> Please take this up with Uncle Ben and the rest and let me know just as soon as possible what you think . . . I believe this is the best proposition we have ever run across and we are making a big mistake if we do not take it up. If you do not like the plan of organization, please make suggestions that are preferable.(21)

In Stockton, secretary George Dickenson believed this represented a good buy—no mistake could be made by purchasing property worth twice the amount paid for it. On October 9, C. Parker Holt wired Pliny, "Have you closed and secured Peoria deal? Count on my support." Soon after, C. Parker wrote that he was convinced that the Caterpillar business was going to be a great moneymaker. Getting the factory immediately was important, he felt, because every day's delay in building engines was just so much money lost. He promised to tell George Dickenson to draw on resources in Stockton for $10,000.(22)

In Minneapolis, Pliny Holt's men began loading flat cars with manufacturing materials for shipment to Peoria. On October 25, 1909, when the court order releasing all rights of the Colean Company was delivered, Baker wired Pliny Holt: "Sale and settlement confirmed by the court today." The Holt Caterpillar crawler industry found a new home in Peoria. (23)

In retrospect, it is surprising that negotiations for the Colean factory were stalled so long over such small sums of money. A major road-block in the negotiations, for example, occurred when the price of the certificate of purchase was raised from $10,000 to $11,500. It is also surprising that the home office in Stockton was so reluctant to come forward with financial support. Not until the deal had been closed was there an offer of aid, and this was one which was limited to $10,000 with no additional assistance to follow.

This record of events suggests that Pliny E. Holt was the prime mover in establishing a Midwestern factory. He made the initial decision to go ahead with this project and monitored the business details. Murray M. Baker was also crucial in these negotiations, keeping the Holt officials informed about the proceedings in Peoria. His acumen and shrewd gambling instincts led him to abandon his own farm implement dealership and affiliate with the new company which he helped to develop. Ben C. Holt in his position in Walla Walla, Washington provided valuable advice in financial matters and C. Parker Holt in Stockton gave solid support to this new business venture.

Meanwhile, others in the home office in Stockton such as Benjamin Holt, George L. Dickenson, R. S. Springer and lawyer, C. L. Neumiller, exerted only minor influence in these important decisions. After the purchase of the property, however, they acted immediately to promote the business opportunities in Peoria. Benjamin Holt, now 61 years of age, planned a trip to the East with stops in Washington, D.C., New York, Concord, New Hampshire, and Peoria, and he firmly requested a family conference in New York on December 19. On December 6, C. Parker wrote to Pliny that he had received a telegram from Benjamin Holt to confirm this conference and added, "As we cannot definitely decide the matter regarding the Peoria proposition until after our conference with him (Benjamin), we will go to New York and wait for him at the Waldorf Hotel." (24)

This rendezvous occurred as scheduled with Benjamin, his wife, Anna, and their son William K. arriving from Washington, D.C., and C. Parker Holt, George L. Dickenson, and Pliny Holt arriving from Stockton and Minneapolis. In their sessions, organizational matters were hammered into shape and a general agreement spelled out in the charter of incorporation, which was filed with the Secretary of State in Illinois on January 12, 1910. The new Peoria officers were Pliny E. Holt, president, George L. Dickenson, secretary, and Benjamin Holt, Ben C. Holt, and Murray M. Baker as additional directors. Pliny, Benjamin, and George Dickenson each received 500 shares of stock at a par value of $100 per share, while long-time company employees Daniel Gilmore and Charles R. Weston each received one share. Capital stock was issued at $500,000, with $230,300 paid up. For the new company's name, the group decided on The Holt Caterpillar Company, not The Peoria Holt Company as had been suggested by C. Parker Holt. (25)

As members of the Holt fraternity greeted the New Year 1910, there was much to be cheerful about. In the past year, the company had made a huge sale of tractors to the City of Los Angeles, and acquired the Best factory in San Leandro, a move which was proving profitable. The demand for combined harvesters remained strong, and the Illinois factory had required very little cash outlay. Dickenson reported that sales in the Stockton office during December of 1909 had risen to $77,910, and Ben C. Holt enthused that harvesters and tractors were selling well in the Pacific Northwest. Nothing had ever been devised before which did such good work, he concluded, predicting 1910 to be "the greatest year in our history."(26)

Chapter 12

CORPORATE GROWING PAINS IN THE MIDWEST

The Holt family's euphoria about its new business enterprise proved short-lived. High expectations descended as 1910 wore on, making it a year of despair. The headaches of managing an enterprise 2000 miles across the nation became acute. Minor weaknesses at every level of operation created major problems in the areas of management, engineering, and finance. Establishing new factories was one thing; keeping them solvent something else.

The expansion of the Holt Manufacturing Company from a regional corporation to a national one changed the character of the firm and created unexpected strains. The Holt Company had been on the West Coast for over 25 years. It was familiar with western farming conditions and knew how to satisfy its customers. It had expanded and bought out regional competitors so that it was strong enough to match any would-be competition in the area.

Doing business in the Midwest and Canada created countless new problems. The company soon discovered, for instance, that it was extremely difficult to sell the new crawler tractors because farmers were accustomed to wheels and saw no good reason to change to track-type treads. This inbred suspicion was exacerbated by the failure of the Holt tractors to arrive in time to be entered in the famous Winnipeg Plowing Trials of 1908 and 1909, where Midwestern farmers could have seen the Holt crawler tractor in action.

Furthermore, the confidence of the Midwestern farmers had been preempted by large and powerful farm implement companies such as the J. I. Case Company (which had sold 8,000 farm engines between 1906 and 1910), Hart-Parr, Rumely, Avery, and International Harvester. Most farmers had already developed a loyalty to one of these well-established firms and their products. In small towns, these companies' dealers had already worked to gain farmers' trust.

Complicating matters, the Stockton firm had not yet established credit in the Midwest. Ben C. Holt complained that he could not borrow money from midwestern bankers because he represented a Pacific Coast company. Short on capital, the Caterpillar Company could only sell its tractors for cash, and this policy antagonized farmers who had long been able to buy farm machinery on credit. The Funk brothers of Bloomington, Illinois, demanded that their tractor, the first produced by the Holt Company, prove itself hauling manure before they would lay down money for the purchase price. In 1947, Murray Baker recalled, "The first tractor we manufactured in Peoria we had to sell for cash in order to have enough money to pay for the building of the second tractor. It was tough going in those early days, mighty tough indeed." (1)

From the beginning, Pliny Holt and his chief engineer, Robert Gotshall, believed it imperative to alter the design of the California Caterpillar track-type tractor to make it more acceptable to midwestern clients. Accordingly, they abandoned the single front steering wheel and attached two front wheels to give the tractor a more conventional appearance. In addition, they added a drive pulley for belt work, adopted a Remy high-speed magneto, installed new timing gears, redesigned the radiator assembly, and increased the speed of the tractors from 2 to 2 1/2 miles per hour.

So many changes were made in the Peoria tractors that it was quickly impossible to get proper repair parts from Stockton where the principal foundry was still based. Because of the paucity of repair parts, Stockton factory manager R. S. Springer refused to permit several Peoria tractors to be entered in the Minnesota and Montana state fairs in 1910, an action which infuriated Pliny Holt. Pliny thought his tractor models were superior to those manufactured in Stockton, and when a Holt engineer in California listed 12 changes which should be incorporated in the design of the Peoria engines, Pliny read the suggestions and tersely jotted down a dozen times, "All right as is."(2)

To accommodate Midwest farmers who were accustomed to tractors with two front wheels, the Holt Caterpillar Company of Peoria, Illinois built a few track-type tractors with two front wheels around 1910.

Machine shop of the Holt Manufacturing Company in Stockton May 14, 1911. Factory featured power transmitted with line shafts and belts.

The Holt companies' worst financial crisis occurred in 1910, just after purchase of the Peoria plant. Although the Stockton factory had returned a profit in 1909, there had been no funds earmarked to finance the Peoria enterprise. To expect the Peoria factory to manufacture and sell enough tractors to meet obligations in 1910 without any outside support was unrealistic, a serious miscalculation that almost ruined the fledgling franchise. Commenting years later on the pressure faced in these early years, Murray Baker wrote: "Neither Pliny nor I ever received a thin dime for our efforts. We had no money for salaries and for our own expenses . . ."(3)

Perhaps the financial difficulties should not have come as a surprise. In July of 1909 Ben C. wrote Pliny to remember that the farm machinery business was seasonal—the company borrowed heavily from banks to build machinery which was sold during the farmers' planting season but not paid for until the harvest was completed. Cash flow thus depended upon crop conditions and market prices. Growing increasingly apprehensive about the proposed Peoria venture, Ben C. grumbled to Pliny in August that it was all right for people to talk about raising capital to start a new business, but it was not so easy to get them to put up actual cash. In October, Ben C. wrote that his associates were too optimistic about the projected profits from the newly purchased Best subsidiary, and in November he said he was tiring of seeing money always going out and little coming in. Although he liked the idea of a Midwestern plant, he could not see where the money would come from to meet the first payment. (4)

Ben C.'s dire predictions came true with a vengeance. In September 1910, the Dime Savings & Trust Company warned that $48,000 would fall due as partial payment for the Colean factory. Ben C. begged Pliny not to run up bills because they would be difficult to pay.

In Stockton, C. Parker likewise deplored the financial crisis, saying he had been fighting tooth and nail to balance his books. Financial obligations to three banks totalled $152,500, and he had only $45,000 on hand. In addition, he needed $12,000 to pay the operating costs of the San Francisco office; $20,000 for the Stockton plant payroll; $15,000 for freight bills; and $5,000 for other bills. In December, expenses would total $100,500, with $27,500 owed to Daniel Best as partial payment for the San Leandro factory. (5)

Pliny added to the gloom in 1910 by requesting that Ben C. in Walla Walla ask C. Parker Holt in Stockton for financial aid. Pliny did so because two Holt tractors in Canada had remained unsold because of crop failures, and the Northern Holt Company had done practically no business in Canada where the economy was sluggish. (6)

Some of the new operation's problems were the result of selling tractors over a wide territory where it was difficult to keep in touch with customers. By 1910, Caterpillar crawlers were in Louisiana, California, Mexico, Idaho, and Canada, and the distances from the home factories created such inconveniences for salesmen, dealers, and repair experts that George L. Dickenson facetiously advised Pliny Holt not to sell any tractors outside a ten-mile radius of Peoria.

Eagerness for business repeatedly lured the Peoria company into making sales beyond a sensible geographical area. For example, in 1909 it sold an engine to John Corrigan of Saskatchewan. A few months later Corrigan complained to Peoria that the tractor was worthless. In response, the company sent representative H. L. Freeman to Canada to straighten out the matter. Freeman's letters over the next several months read like the lamentations of Jeremiah. First, Freeman could not find the engine, going to Regina, then to Moose Jaw, and finally to Rouleau. There he rented a livery rig and drove nine miles into the country, where he discovered the tractor with a broken oil pump, a cracked flywheel, two burned-out bearings, several broken piston rings, a scorched cylinder, a missing fan, cracked governor gears, and a badly leaking radiator. He immediately wired for repair parts, but these were held up by custom officials at the border. (7) Meanwhile, Corrigan insisted he would not keep the tractor under any circumstances. He hired a lawyer and said he could "law as long as they could." At last, Freeman borrowed $75 in order to return to the United States, suggesting that Pliny Holt might himself visit the angry farmer. When Corrigan was finally advised that the company would replace his tractor with a new one, the disgruntled man wired, "I don't want the one I have so I cannot understand why I should want another one. I want you to take this one back. It is no good."(8)

This episode illustrated another weakness in the early Peoria organization: the absence of a good system of dealers, salesmen, and expert road men to provide needed services for individual tractor owners. Pliny initially believed that these kinds of employees were unnecessary because people would come to the factory anyway, but Ben C. insisted that the Peoria office had to start advertising and sending salesmen over the territory because "You can't do business in a vacuum. We will have to get out and hustle like all the other tractor people."(9)

Traveling salesmen, Pliny learned, played a crucial role in the distribution of farm machinery. Valise in hand, these "knights of the grip" swarmed over the countryside to canvass prospective customers. Living a rugged life, they traveled in all kinds of weather, catching midnight trains and riding slow freights. Often a salesman stopped at a hotel only to learn that a rival had registered ahead of him and already had his prospect's order in hand. Occasionally a smooth agent would attend a local church on Sunday and offer to give the morning sermon. After unloading some fire and brimstone he would conclude with an invitation to the faithful to buy his tractors. (10)

At the Holt Manufacturing Company, Dan N. Gilmore became the top salesman, while in the Best organization Fred Grimsley was equally indefatigable, often rising at four o'clock in the morning to begin his search for prospective buyers. (11)

In addition to sales work, many road men made collections as well. Farmers usually made down payments and signed promissory notes which fell due in one to three years. Although the Holt Caterpillar Company in Peoria had originally sold its tractors for cash only, it was soon forced to extend credit in order to match the practices of its competitors. Sometimes collections were much more difficult than sales. C. Parker Holt said in 1910 that he was unable to pay his creditors because he had been unable to collect from farmers. In Walla Walla, Ben C. lamented that even though he had tried every known means to collect from farmers, he had meager results because crop yields and prices were low. Writing to Pliny, he gloomly remarked, "I get discouraged and would go and jump in Mill Creek, if it were not for the fact that there is no water in it." (12)

A few disreputable farmers bought machinery with no intention of paying for it. Some defaulted on payments and let the manufacturer repossess the machines. Others left the state to avoid debts, remaining one jump ahead of the sheriff, while still others registered their property in their wife's name to prevent foreclosure proceedings. One farmer said he received so many bills in the mail that he needed a special sack to carry them home where they were stacked up like cord wood. A collector for the Colean Company in 1908 complained that one farmer was the devil and then some. He refused to pay the note and told the agent to go to a place reputed to be warmer than this one. One diligent collector waited until his client went swimming and then had the sheriff grab the swimmer's clothes, gold watch, and chewing tobacco, which were sold at auction for $40. Another collector refused to be bonded, claiming that when a man had finally succeeded in making a collection, he would be too tired to steal. (13)

In addition to monetary problems at the new Holt Caterpillar Company in Peoria, Pliny Holt also faced new engineering problems. First, the tractors manufactured in California proved inadequate for use in the Midwest, where tough sod and hard clay soils required heavier engines. Writing to Ben C. in 1910, Pliny confided he had lost faith in the Stockton tractors and would have to rebuild the whole bunch, a view which had been confirmed by other engineers such as Henry Preble and Paul Weston. But Pliny found himself opposing Benjamin Holt on this matter, and while Pliny was unhappy about manufacturing tractors which he did not think could be successful in the Midwest, he did not have the authority to make such a radical change.

In the meantime, for his part, Benjamin Holt wanted to begin manufacturing a small crawler tractor to sell for about $2,000, suitable for the average farmer. He also advocated building a wheel tractor for the Midwestern trade. While these were innovative ideas, Pliny believed it would take too much time to test the new models because they had already spent almost two years in Minneapolis and Peoria without reaching full production.

Addressing the conflict between Benjamin and Pliny, Ben C. Holt advised Pliny to be more independent, because he knew the Midwest better than anyone else in the company and therefore should forge ahead and ask questions afterwards. He suggested, however, that Pliny would do well to accommodate Benjamin until the Peoria plant was on its feet, and then he would no longer have to make compromises or humor Uncle Ben to maintain peace in the family: "Pat him on the back, compliment him, and stay out of arguments, because harmony is extremely important." (14)

Adding to Pliny's worries was payment of $44,000 for the Colean factory due on November 13, 1910. He thought that new capital to meet this note could be raised by selling bonds backed by the Holt Caterpillar Company of Peoria, but the floating of these bonds was delayed due to a misunderstanding between C. Parker and Pliny.

C. Parker neglected to facilitate the stock sale, causing Pliny to believe that C. Parker had deliberately turned against him. C. Parker, on the other hand, seemed to be critical of Pliny's repeated requests for financial assistance. On one occasion, C. Parker wired, "Go out and sell some tractors and you will have plenty of money." On another, he wrote: "You run around too much . . . Stay home and tend to your business and things will improve."(15)

That some friction occasionally developed among the Holts is not surprising, and perhaps the most remarkable thing about the Holt management over the years was the near absence of feuds among the families. Over a long span, their high degree of cooperation proved to be one of the keys of their success.

As Pliny Holt continued at his post in Peoria in 1910, bills piled up, and engineering details cried out for attention. Then his health began to fail. Learning that Pliny was thinking of resigning, Ben C. wrote a series of letters urging him to stay on the job. He said all the Holts were having a tough time and that Benjamin thought more of him than all the rest of the clan put together. Ben C. concluded, "I couldn't make one gear in the Caterpillar, yet if I had your noodle, I would consider my assets had enhanced about 5000 percent, so for goodness sake don't get discouraged and go West and live on a farm . . . Remember that I am with you heart and soul."(16)

After months of weighing matters, Pliny resigned in November 1910. He turned the management of the Peoria plant over to George Dickenson, Murray Baker, and Fred Moore, a Stockton plant engineer. (17)

By 1912, Pliny's health improved and he re-assumed executive duties with the Holt Manufacturing Company, moving on to a brilliant engineering career. Likewise, the Peoria factory got on its feet after a shaky start and began moving forward into an exciting era. But the growing pains, vexing problems, mental anxiety, and the mountain of work that made it happen were not forgotten.

Chapter 13

THE HOLTS REGROUP ON THE EVE OF WORLD WAR I

On an autumn day in 1910, C. Parker Holt approached Benjamin Holt with the idea that they should try to increase tractor sales in foreign countries, especially in South America. Although Uncle Ben himself had taken his combined harvester to Australia and had his machines demonstrated in Spain, he was feeling the pressure of the Holt business crisis and he summarily told C. Parker to drop the idea of exploring the South American market. There was plenty of work to be done at home.

Taking the initiative, C. Parker quietly shipped a Caterpillar crawler and combine to Buenos Aires, where he demonstrated the oufit for several months for enthusiastic growers. In March 1911, he cabled Benjamin to ship nine Holt tractors to a farm machinery dealer in Buenos Aires, a $40,000 order that was just the beginning. When he returned to Stockton, he carried another sheaf of orders into Ben's office.(1)

Sophisticated and well-qualified to promote Holt business abroad, C.Parker had a good background in engineering theory from the University of California and Cornell University, practical experience from working in the factory shops in Stockton, and a knowledge of farming from the Holt 1,000 acre ranch in the San Joaquin Delta. Agricultural historian F. Hal Higgins doubted that any man since Cyrus McCormick had contributed so much to the successful exportation of American farm machinery.(2)

For the Holt company, C. Parker's foray into South America was tantamount to discovering gold. Argentine farmers, dissatisfied with European steam traction engines, eagerly began purchasing the Holt tracklaying machines. In June 1911, Superintendent R.S. Springer wired Secretary George L. Dickenson stating that they had received orders for ten more tractors, making a total of 32 South American sales. Before the end of the year, 134 gasoline tractors had been shipped to Argentina, almost half of the company's total tractor sales in 1911.(3) This new export business led executives to transfer the export trade from Stockton to Peoria, a move which would reduce freight costs to most foreign markets. Ben C. Holt was placed in charge of the new export office in New York City in 1915, where he handled business with 45 countries.

After the difficult year of 1910, foreign and domestic sales climbed. By 1912, business had improved with a production of 306 tractors compared with 105 in 1910, and the Holt Manufacturing Company total sales reached $2,232,000. The number of employees in the factories in Stockton, Peoria, Walla Walla and San

Holt Caterpillar crawler wins prize for tractor performance in Argentina in 1912.

Leandro reached 900 workers with a monthly payroll of $35,000.(4)

Despite these favorable financial developments, the management of this sprawling conglomerate was under increasing strain because its unwieldy business structure was inefficient and even chaotic. During the founding of the Peoria factory, for example, executive directives lacked clear lines of authority. In addition, although Benjamin Holt was president of the company, he was largely preoccupied with engineering matters. Secretary George L. Dickenson not only managed the Stockton home office, but was also expected to supervise policies in the former Best factory in San Leandro where Clarence Leo Best was nominal head of the Holt subsidiary. In 1910, Best resigned. He claimed his authority had been undermined by San Leandro plant secretary O.F. Meyers, who was, without Best's knowledge, given powers normally reserved for the president, including the signing of checks.(5)

Holt officials in Stockton, on the other hand, believed Best's resignation was inevitable because he had never favored the sale of his father's firm to the Holts. In fact, they were not even surprised that he established his own factory in Elmhurst, where he began manufacturing tractors in direct competition with the Holt Company.

Through the rapid acquisition of competitors and branch offices, the Holt management had become extremely fragmented. In 1911, C. Parker Holt was demonstrating Holt machines in South America and Pliny Holt was recuperating from illness, while other family members and officials were responsible for the branch offices in Los Angeles, Denver, Portland, Walla Walla, San Francisco, and Calgary. Also included in the Holt stable of enterprises in these years were eight legally separate companies: The Holt Manufacturing Company, Aurora Engine Company, Houser-Haines Company, Stockton Realty Company (all of Stockton), Holt Caterpillar Company of Peoria, Canadian Holt Company of Calgary, Northwest Harvester Company of Spokane, and Best Manufacturing Company of San Leandro.

Not only was the business structure of these interwoven operations complex, but crucial decisions regarding direction and finances were faced on every hand. Unwise actions by company managers could result in a corporation being taken over by bankers if they picked up a controlling interest in the stock. Finally, although each of the Holt enterprises had proved profitable in the past, there was no guarantee that they would continue to be so in the future.

At this critical juncture, Charles L. Neumiller, long-time legal counsel for the company and a close friend of Benjamin Holt, exerted a timely influence. Keenly aware of managerial problems, Neumiller concluded that the far-flung Holt business should be reorganized to form one large corporation. This meant the dissolution of all seven subsidiaries and a conversion of their stock into shares of the new Holt Manufacturing Company.

THE VANCOUVER WORLD Saturday, March 2, 1912

THE MOST POWERFUL TRACTION ENGINE IN THE WORLD.

CATERPILLAR IS POWERFUL CONTRIVANCE

Modern Mechanical Device for

necessity of these mechanical power appliances, when it comes down to scientific farming.

Farming in these days is not done in the haphazard fashion of former years. Everything is figured out to a nicety and if the weather is at all lenient in the matter of frosts, bumper crops are invariably the result of the farmer's labors. Nine out of ten farmers ... that the results achieved by them is brought about by the use of the most modern methods in the cultivation of the soil. Hardly a farm is seen on the prairies without its mechanical equipment. The tillers of the soil have come to learn that ...

where a horse cannot go and will ... ditches that a member of the ... breed could not overcome ... ning lose. One of the many ... of the Caterpillar is that it ... track and pulls one up again.

It crawls over the softest ... and reclaimed land and will ... furrows at a stretch on land where a horse will break through and ... down.

This powerful gasoline or steam apparatus will drive a threshing ... ine, pump water, saw wood ... steep inclines, plowing at the same time if required, haul heavy ...

On March 2, 1912, "The Vancouver World" hailed the Holt track-type tractor as "the most powerful traction engine in the world".

Although this drastic action might not please all the individual stockholders, this merger, according to Neumiller, was imperative.(6)

Benjamin's son, William K. Holt, remembers that this major corporate reorganization was accomplished because "Charlie Neumiller simply laid down the law. This was the course of action and all the Holts had to fall in line." On December 31, 1912, the eight companies were merged into one giant corporation with the Holt family holding the controlling stock. This dramatic move kept the heavily debt-financed company operations afloat for the next decade.(7)

At the time of this legal consolidation in 1912, Benjamin Holt was the last surviving Holt brother who had moved from New Hampshire to California. A. Frank had died in 1889, William Harrison in 1904, Charles Henry in 1905, and the youngest brother Fred Samuel in 1909. Uncle Ben owned the largest share of the corporation. He also drew an annual salary of $12,000, and during the consolidation he received $200,000 for his 21 patents.

Although there are no papers that show how the total reorganized company assets of $2,208,000 were divided among the owners, records indicate that by mid-1913 each of the following men owned $100,000 worth of stock in the new Holt Manufacturing Company: Pliny, C. Parker, Ben C., son Edgar Holt; R.S. Springer, George Dickenson.

After reorganization, the Holt Company entered a new period of rapid growth. The first big spur to sales came from an unexpected source—a challenge by a newspaper editor in the "New York Evening Journal". Arthur Brisbane owned a large New Jersey farm, and on July 1, 1913, he asked: "Has anybody yet made a motor to do farm work?" Looking for a machine that could uproot trees, plow the soil, and furnish belt power for threshing grain, he invited manufacturers to demonstrate their tractors on his 1,000-acre farm. Quickly recognizing a chance for free advertising and exposure on the East Coast, the Holt Manufacturing Company alone accepted the challenge and shipped Brisbane a tractor.(9) In a second editorial, Brisbane extolled the virtues of the Holt crawler tractor, which crossed soft ground, toppled trees, and pulled ten plows cutting furrows 14 inches deep. He concluded, "If you want to see the most wonderful "caterpillar" that you have ever heard of, the most interesting monster in or out of Holy Writ, go down to Farmingdale, New Jersey, and ask where the Holt Caterpillar is working."(10)

Picking up on Brisbane's enthusiasm, the "New York Tribune" and the "Christian Science Monitor" claimed the Holt machine was uprooting trees 50 feet tall in less than a minute's time, while the "Byron Times" in California ran banner lines, "Holt Caterpillar World Marvel." These unsolicited testimonials no doubt had more publicity value than all the paid advertising issued by the company that year.

Convinced of the immense value of advertising in making Holt machines known outside the West, Benjamin Holt himself decided to do some promoting. He and his son, William K., traveled to Russia in 1913 to demonstrate the company's self-propelled combined harvester. When they arrived in Kiev, they were dismayed to see that their agent had been unsuccessfully trying to harvest damp and green grain. To correct the negative impression this was creating, they shipped the harvester south to the Black Sea where the grain stood dry and dead ripe and the machine predictably did excellent work.

In the same year, the Holts hired their first advertising manager, L. W. Ellis, to promote track-type machines and produce a monthly house organ. Ellis re-designed the Caterpillar logo into a wavy line of letters reflecting the movements of a caterpillar.

A 1915 Caterpillar advertisement.

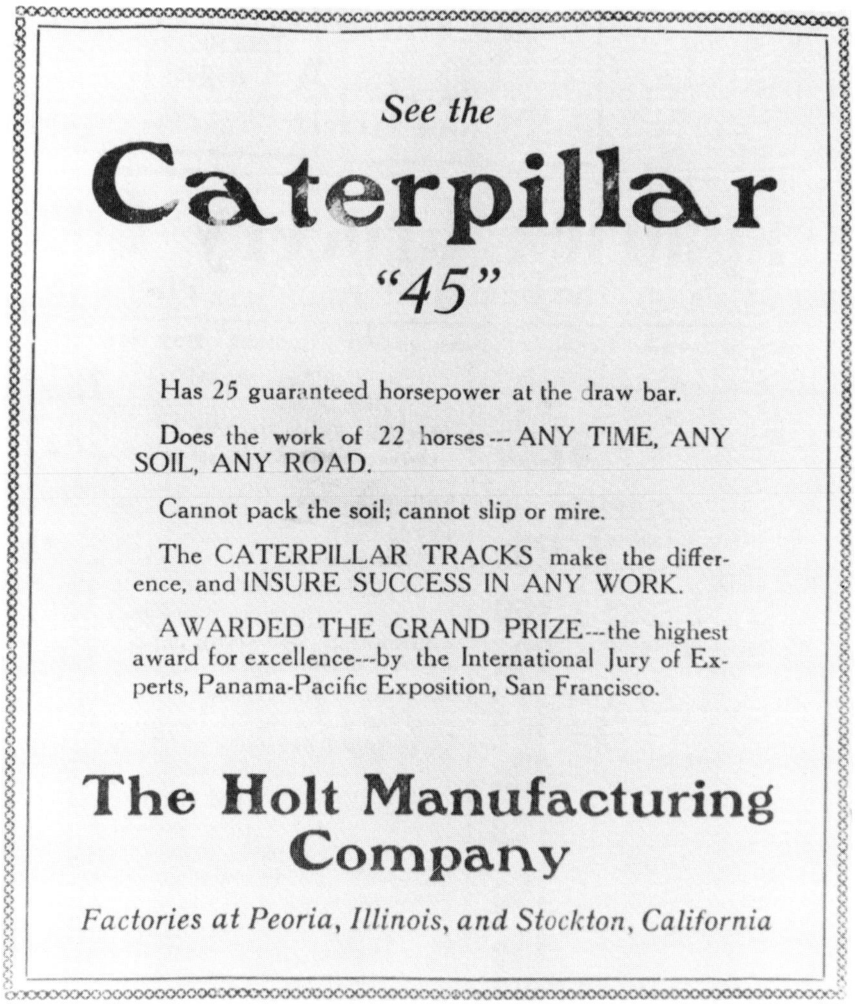

See the

Caterpillar

"45"

Has 25 guaranteed horsepower at the draw bar.

Does the work of 22 horses --- ANY TIME, ANY SOIL, ANY ROAD.

Cannot pack the soil; cannot slip or mire.

The CATERPILLAR TRACKS make the difference, and INSURE SUCCESS IN ANY WORK.

AWARDED THE GRAND PRIZE---the highest award for excellence---by the International Jury of Experts, Panama-Pacific Exposition, San Francisco.

The Holt Manufacturing Company

Factories at Peoria, Illinois, and Stockton, California

In 1914, additional attention was forced on the Caterpillar crawler during a well-publicized plowing demonstration near Fremont, Nebraska. In that contest one of the Holt engines set a world's record by pulling 24, 14-inch plows, cutting a strip 28 feet wide and 7-1/2 inches deep in hard, dry "gumbo" soil. Experts estimated that this feat required 115 horse-power.(11)

Impressed with Holt's growing reputation for producing heavy-duty, high-performance vehicles, the Southern Pacific and Santa Fe Railroads began using Holt tractors for hauling freight to various terminals, while Nevada mining companies ordered them for hauling ore cars. One Arizona contractor on a canal project tried moving dirt with a 40-80 Flour City tractor, and a 30-60 Aultman Taylor tractor, but both failed to do the job. As a last resort he secured a Holt tractor which completed the job without difficulty. The "Oregonian" reported in 1912 that Caterpillar crawlers were pulling wagon trains loaded with 15 tons on 150-mile trips in the Northwest and one tractor was doing the same on the Yukon route near White Horse in Alaska. A Holt advertisement in the "Country Gentleman" featured this account of the Yukon tractor's exploits:

> Forty below in Alaska—nothing stirring but dog teams and the Caterpillar tractor. When the boats and trains of the White Pass and Yukon Route quit for the winter, the Caterpillar began covering a sixty-mile haul on the ice with fifty tons of perishable foodstuffs. Motor housed in ten big sleighs and a caboose behind, the tractor with a double crew bucked the drifts twenty-four hours a day and fed the camps up river . . . (12)

When the automobile reached rural America it created a demand for better roads. The early motorists routinely traversed dusty roads in dry weather and faced quagmires in the rainy season. As these drivers negotiated deep ruts, tree stumps, and snow banks, they increasingly endorsed the Good Roads Movement, a crusade to encourage government officials to allocate more money for highway construction.

Meanwhile, it became evident that the powerful Holt tractors were particularly suited for grading roads because of their superb traction and maneuverability in rough terrain. At ease pulling road graders and hauling freight, as well as plowing fields, track-type tractor were only beginning to reach their full potential.

Pliny Holt is shown with his own home-built automobile in Stockton, around 1907.

Chapter 14

TRACTORS AND TANKS IN WORLD WAR I

It was the theaters of War in Mexico and Europe, however, which proved a great challenge to the management of the Holt factories in Peoria and Stockton. Using farm tractors for pulling vehicles was one matter, but turning them into instruments of warfare was another. This process of technological evolution from the tractor to the military tank proved to be one of the most significant developments of World War I.

Since the manufacture of motorized automobiles and trucks around the turn of the century, military experts had debated whether or not animals provided the best form of transportation in war time. Some experts insisted mules were preferable to trucks because mules could pull supply trains over rugged ground for three days with nothing but wheat straw in their bellies, while motorized vehicles needed constant supplies of fuel. Officials in the United States Army firmly believed that trucks could not be used on steep grades or on mud roads, and prior to 1912, they had purchased only 28 motor trucks, despite knowledge that the French and Russian governments were buying as many as 125 from American manufacturers in a single order. After conducting maneuvers with mules and trucks in the Midwest in the summer of 1912, the military officials summarily concluded, "At present no three-ton truck can be depended upon to transport army supplies over country roads."(1)

Holt 10-ton crawler tractor hauling supplies to support the United States Army expedition into Mexico.

Officials at the Holt Manufacturing Company were of a different mind. In November 1913, Ben C. wrote Senator Miles Poindexter, chairman of the Senate Committee on Expenditures in the War Department, that a special track-type tractor would enable the War Department to reduce the number of horses used in wartime. Holt added, "We have already shipped some of these machines to foreign lands, and we are meeting with more or less success in interesting various war departments in Europe in this type of Caterpillar." In conclusion, he wrote that he would gladly demonstrate this tractor for officers of the War Department. (2)

Receiving no reply, Ben C. sent a second letter to the War Department Board of Ordnance and Fortifications, explaining that he was mailing an issue of the "Caterpillar Bulletin" which described the tractor's ability to move over soft ground, and he wanted to demonstrate this tractor to military experts. A curt reply from the captain of the Corps of Engineers on December 20, 1913, said it was uncertain when the Ordnance Board would meet to discuss these matters. Meanwhile, Senator Poindexter forwarded Ben C.'s letter to Major Leonard Wood, president of the Board of Ordnance and Fortifications, who showed no hurry in taking any action despite the fact that Europe was on the brink of war. (3)

Irked by these delays, Ben C. again wrote to the War Department on May 11, 1914, with the offer that Murray Baker, vice-president of the Holt Caterpillar Company of Peoria, would demonstrate the track-type tractor at any designated place without expense to the federal government. This demonstration would prove that there was absolutely nothing like this high-power tractor for working in all soil conditions, year in and year out. (4)

Again the military showed no interest, a policy which continued even after the governments of Great Britain, France, and Russia began purchasing Holt tractors for the Allied cause. It was not until May 1915 that the War Department gave the Holt Company permission to demonstrate a 75-horsepower Caterpillar crawler at the Rock Island Arsenal in Illinois, requiring that the company pay for the cost of these experimental trials.(5)

The Rock Island tests proved that the Caterpillar crawlers could haul heavy artillery and ammunition through mud where neither horses nor any other known motive power could do the job. The editor of the "Field Artillery Journal" estimated that the cost of supporting mules was three times more than tractors and, in addition, tractors required less space on the road, needed less care, and permitted armies to operate farther from their base of supply. (6)

When additional tests at Fort Sill, Oklahoma, in November 1915 proved successful, the military decided to eliminate animals from the Ninth Field Artillery stationed in Honolulu—making it the first completely motorized artillery regiment in the army. In September 1916, the "New York Times" claimed that this would be the first horseless regiment in the world: horsemen would ride on motorcycles while tractors would move weapons and supplies. When the Bureau of Ordnance subsequently let competitive bids for 27 track-type tractors, the only bidder was the Holt Manufacturing Company of Peoria. (7)

Holt Caterpillar "45" in severe swamp land tests before the US Field Artillery Board at Fort Sill, Oklahoma around 1912.

Holt track-type tractor pulling artillery during the Battle of the Somme in the summer of 1916.

Military experts received an excellent opportunity to observe the performance of Caterpillar track-type tractors during the United States invasion of Mexico in 1916. This punitive expedition resulted from a series of clashes between the two countries brought about by unstable political conditions in Mexico and American interference in the country's domestic affairs. When President Woodrow Wilson sided with the political faction led by Venustiano Carranza, the infuriated rival faction led by Francisco Villa raided Columbus, New Mexico, burning the town and killing 19 inhabitants. In response, Wilson ordered Brigadier General John J. Pershing to invade Mexico and capture Villa. The American Army crossed the border on April 8, 1916, and pushed southward 350 miles to Parral without capturing the Mexican leader. Wishing to avoid further military involvement in Mexico, especially while war was raging in Europe, President Wilson ordered the withdrawal of American troops from the Mexican side of the Rio Grande on January 27, 1917, ending the fiasco. (8)

Although "Blackjack" Pershing never spotted the elusive Villa in the deserts of Chihuahua, Pershing's expedition did serve as a proving ground for new American trucks and tractors. For the first time, an American military force moved into enemy territory with its entire supply line dependent on mechanized units. Crawler tractors built roads across the desert and carried fuel, water, ammunition, and rations over sand dunes and rocks and through treacherous arroyos during vicious sand and rain storms. One tractor hauled as much freight as 20 motor trucks, and track-type tractors reportedly moved freight at one-fifth the cost of mule trains. Pershing claimed that the proven perfection of the gas engine was the greatest of all discoveries in this campaign in which efficient motor vehicles increased the scope of military action. He added, "The expedition into Mexico would have been impossible without the tractor and motor truck."(9)

However, the evolution of the farm track-type tractor into a military vehicle had earlier beginnings. The story begins in 1910 when Dr. Leo Steiner, a Hungarian engineer and farmer who grew dissatisifed with his steam traction engine, ordered a Holt tractor from Stockton. Highly pleased with the machine, he acquired a Holt agency in July 1912, and was granted exclusive rights for sales in Austria-Hungary and Germany. (10)

Meanwhile, officials of the Austro-Hungarian War Department asked Steiner to conduct secret field trials testing the Caterpillar crawler against several European tractors in pulling huge Austrian howitzers across loose sandy soil and marsh lands. In October 1912, Steiner's Holt tractor easily out-pulled a 120-horsepower Porsche tractor in a demonstration outside Vienna. Another exhibition took place near Magdeburg, Germany, but German military experts erroneously concluded that the engine was "of no military importance." They would later find that this disinterest in military track-type vehicles was one of their fatal errors of judgment.(11)

During the third trial in Austria in May 1914, a Porsche tractor pulled a howitzer 500 feet into a swamp and then sank into the mud, while an English tractor also failed. The Caterpillar crawler, on the other hand, moved the gun 250 feet farther than the Porsche and then, as the driver raced the engine and threw in the clutch, it dragged the heavy gun out of the swamp. Steiner later recalled:

> I stood nearby. I heard a shot-like crack, saw the Holt rear up and almost somersault. In the next moment, I saw it return to a horizontal position, lift the gun wheels out of the ground and haul it away. Troops cheered and members of the military committee waved their hats and ran to offer congratulations.(12)

The Austrians were so impressed that they arranged with Holt officials to build a factory to produce track-type tractors. When war broke out in 1914, however, shipments of materials from Holt factories in the United States were terminated. Subsequently, the Central Powers' military leaders ignored the track-type tanks as engines of war until it was too late. Following the British introduction of military tanks in the Battle of the Somme in September 1916, the Germans attempted to build military tanks on the Caterpillar principle, but they were only able to build about 75 A-F-V tanks before war's end, some 15 of which engaged in battle.(13)

In contrast, Caterpillar crawlers served the Allies well in Western Front trench warfare, where maintenance of supply lines had become crucial. Consequently, officials of the British government invited several Holt engineers who had attended the tractor trials in Austria-Hungary to demonstrate the versatility of their Caterpillar crawlers in England in October 1914. Pleased with the results, the British War Office placed an order through J. P. Morgan & Co. of New York for 10 Caterpillar track-type tractors in March 1915, increasing subsequent orders to 50 and 100 tractors. The French and Russian governments also made purchases, until the total sales of Holt tractors to the Allies in September 1916 numbered 1,000. (14)

The motors for these tractors were manufactured in Stockton and the vehicles assembled in Peoria. It was done in secret because the United States was still neutral, and strong anti-preparedness sentiment existed in the United States. Whenever questioned, the Holts were quick to respond that they sold farm tractors and had nothing to do with putting armor and guns on these vehicles, although the British might be doing something with them other than hauling supplies. By September 1916, 1,200 Holt tractors valued at $3,500,000, had been shipped to England, France, and Russia, according to the "San Francisco Bulletin".(15)

Once unloaded at seaports in France, the movement of these giant tractors to the battle front, by one driver's account, created a buzz of excitement. With big flywheels churning the air, Caterpillar crawlers pulling 14-ton guns, passed through villages crossing bridges so narrow that the engines scraped the fragile iron railings.

Extremes in transport traction contrasted: camels meeting a "Caterpillar" track-type tractor en route—taken aback at the rattle and smell.

As the drivers approached the front, they began moving the big guns into position:

> We slip, we skid, we flounder and stick in soft ground. Day begins to dawn. We race the engine. The good old Caterpillar struggles to get away with its load, but the miry ground is too much for her; the gun sticks in the mud. Get all the boys on the rope. Now are we all ready? Race the engine, let in the clutch, and pull with a mighty heave. All together. Heave again. Hurrah. Up she comes slowly out of the mud. Uncouple the Caterpillar. Turn the gun around and run her under the trees. Some branches lopped from trees cover it . . . It is now daylight but we are safe . . .(16)

As the war continued and a trench warfare stalemate developed in France, British War Department officials began to investigate secret weapons that would repel the opposition's dug-in military forces. What was needed was a machine that could cross trenches and cut through barbed wire entanglements while giving maximum protection to the vehicle operators. While armored cars previously had been used to attack machine gun emplacements, their limited traction made them virtually useless in muddy, shell-pocked terrain.

In the search for a better solution, British engineers first decided to build a 300-ton machine with 40-foot wheels that could roll over trenches, but this project was abandoned in 1915 as too easy a target for enemy gunners. Next they explored using tracklayers similar to those built by the Holt Manufacturing Company, but the machines were plagued with technical problems such as the failure of the tracks to stay on the sprockets of the tractors. (17)

In the meantime, Colonel E. D. Swinton of the Engineering Corps of the British Army had received a letter sent by a friend in July 1914 about "a Yankee machine in Antwerp they call a 'Holt Caterpillar

STOCKTON FILLS WAR ORDERS

HOLT CATERPILLLAR GAS TRACTORS ARE BEING TRANSFORMED INTO ARMORED FIGHTING MACHINES BY THE ALLIES. THEIR SUCCESS IN HURDLING TRENCHES AND SHELL CRATERS IS THE MARVEL OF MODERN WAR HISTORY—THE MOTORS MADE HERE

That the big 120 horsepower Caterpillar tractors being manufactured for the British government by the Holt Manufacturing Company at its plants in Stockton, California, and Peoria, Illinois, are now being used as armored fighting machines in France in recent assaults on German trenches is the news which comes out from Washington today.

The success of the British "tanks," as they have been called, has been such as to attract world wide attention. Until recently these Caterpillars, designed originally for modern farming, have been used by the allies principally to tow big guns. Now, according to dispatches, they have been transformed into fighting machines.

The way they hurdle German trenches, crawl over shell craters and through swamps and marshes is the marvel of the modern age. Except for their armor, their machine guns and their crews, they are no different than the thousands of Caterpillars now in use in farming operations in the United States today.

Stockton is the home of the Caterpillar tractor. It was invented here by Benjamin Holt, president of the Holt Manufacturing Company.

PLINY HOLT TALKS

"We have been shipping two Caterpillars per day to the British government for the past eighteen months," said Pliny E. Holt of Stockton, vice president of the Holt Manufacturing Company, this morning. "All of the motors are manufactured in Stockton. We ship them to our Peoria plant, where the Caterpillars are assembled. We have no direct knowledge of the purpose for which they are to be used. They are shipped as agricultural machinery. Of course, we have representatives over in England, but they are sworn to secrecy. If the Caterpillars are armored, that is done after they reach England. Yes, I should say that we have shipped approximately 1000 of the big tractors to Europe."

The Stockton plant of the Holt company has been working overtime for months, rushing out war orders. The work has been done quietly. Until a dispatch quoting M. M. Baker, vice president of the company in charge of the plant at Peoria, came out from Washington this morning, the fact that the Holts were supplying machines to the allies was not generally known.

Mr. Baker Talks

"We have had nothing to do with putting armor on them or placing machine guns, but some of our men at Aldershot, England, recently were notified that the British government intended to armor some of the tractors and use them for work other than the usual towing of big guns," Mr. Baker was quoted as saying.

Continuing, Mr. Baker said: "Germany had some of these tractors before the war began, and, although I do not understand just how it occurred, I believe she may have got others since then. We have sent some to France and some to Russia. So far as I know, up until the recent appearance of the motor cars, the tractors were used only to tow big guns. I understood that Germany had about 40 of them in this work before Liege early in the war, and recent photographs show that the British are using some of them now for the same purpose."

Mr. Baker said the tractors sent to England weigh about 18,000 pounds each, develop 120 horsepower and are built of steel.

The Operation

The Caterpillar feature, he explained, is of the utmost importance. Speaking broadly, the tractor crawls on two belts, with corrugated surfaces on either side of the body. The corrugated surface is on the ground. On the inside of the belts, on each side of the body, are too lines of steel rails, making four lines in all. These rails are in short sections, jointed and operate over a cogged mechanism that actually lays them down with their belt attachment as the tractor moves ahead and picks them up again so that the car runs on its own selfmade track continuously.

This false war rumor appeared in the "Stockton Daily Evening Works" on September 19, 1916 and in many other newspapers in the United States.

Tractor'. This machine climbs like hell." Swinton later remarked that soldiers had wanted such a machine ever since the time of Julius Caesar. On June 6, 1915, Swinton drafted criteria for designing a successful trench warfare machine: his ideal vehicle could climb a five-foot parapet, travel four miles per hour, cross an eight-foot trench, carry six-pound guns, and be protected with 10-millimeter steel armor. (18)

Given these guidelines, Walter Wilson of the British Foster Company designed a machine on which elliptical tracks were carried up over the high hull of the vehicle. Because this radical rhomboid-shaped machine was eight feet high—too vulnerable to attach a revolving turret on top—guns were mounted in "sponsons" on either side of the hull. Tracks made of pressed steel were attached to this prototype, first called "Landship," then "Centipede," "Little Willie," "Mother," and eventually "Mark I." It was demonstrated successfully on January 16, 1916, the date which now marks the invention of the military tank. (19)

The term "tank" was chosen as a subterfuge so that the machines could be sent into battle as a complete surprise to the enemy. Because the exterior plates formed a box-like container or tank that looked like it could hold water, officials spread rumors that they were designed to haul water in Egypt. When holes were made in the sides for the guns, word was circulated that the machines were revolving snow plows for use in Russia. Signs painted on the machines announced, "With Care to Petrograd." (20)

When tanks were introduced in 1916, they created a sensation on both sides of the battle line. For the first time in history, a machine of great mobility also offered personal protection and fire power, thus providing a unique self-contained offensive unit. One German officer called the new tanks "as cruel as they are effective."(21)

During World War I, the Holt factories at Stockton and Peoria sold 5082 tractors to Great Britain, France, Russia and the United States.

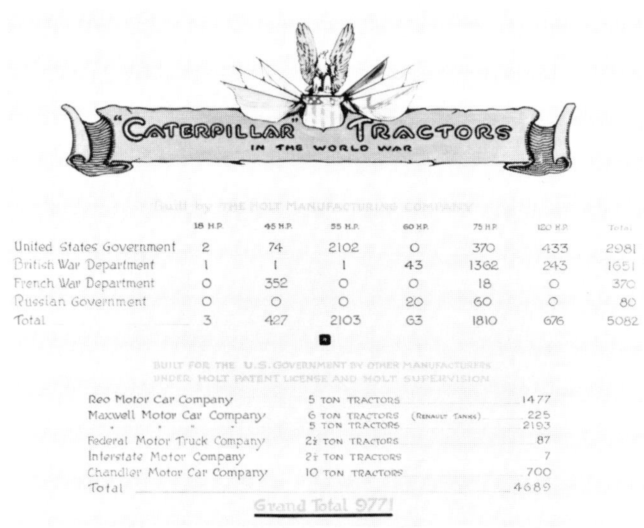

"CATERPILLAR" TRACTORS
IN THE WORLD WAR

Built by THE HOLT MANUFACTURING COMPANY

	18 H.P.	45 H.P.	55 H.P.	60 H.P.	75 H.P.	120 H.P.	Total
United States Government	2	74	2102	0	370	433	2981
British War Department	1	1	1	43	1362	243	1651
French War Department	0	352	0	0	18	0	370
Russian Government	0	0	0	20	60	0	80
Total	3	427	2103	63	1810	676	5082

BUILT FOR THE U.S. GOVERNMENT BY OTHER MANUFACTURERS
UNDER HOLT PATENT LICENSE AND HOLT SUPERVISION

Reo Motor Car Company	5 TON TRACTORS	1477
Maxwell Motor Car Company	6 TON TRACTORS (RENAULT TANKS)	225
" " " "	5 TON TRACTORS	2193
Federal Motor Truck Company	2½ TON TRACTORS	87
Interstate Motor Company	2½ TON TRACTORS	7
Chandler Motor Car Company	10 TON TRACTORS	700
Total		4689

Grand Total 9771

From a military standpoint, however, tanks were only partially successful. Of the 49 tanks assigned to the first action, 17 broke down before reaching the battlefield, nine arrived too late for the fighting, nine never left the starting line because of mechanical problems, five became stuck in the mud, and only nine completed their mission. Inside, the crews faced extreme difficulties, with four men driving the vehicle and four others firing guns. Noise levels were unbearable, and the temperature inside reached 100 degrees. Footing was unsteady in the pitching and rolling cabin, while the prism glass periscopes and open slits gave only a limited view of the battlefield. (22)

After General Douglas Haig announced the use of a new "type of heavy armored car which has proved of considerable utility," Allied journalists embellished the story to suit their imaginations. Edinburgh's "Scotsman" newspaper claimed the new tanks sent the enemy bolting across the battefields like rabbits. One reporter for the London "Times" said the tanks had thrust themselves into enemy lines like hedgehogs into a nest of snakes, while another described the tanks as huge, shapeless hulks, "painted in venomous reptilian colors," that wandered along like some antediluvian brutes which Nature had made and then forgotten.(23)

In the United States a similar journalistic furor followed. Some newspapers carried actual photographs of the first British tanks, while others showed imaginative artists' renditions based on limited information. The "San Francisco Daily News"' picture, allegedly made from an actual photograph, revealed a tank 55 feet long and 10 feet wide. Another paper showed a Cyclopean freak with large revolving knives. (24)

When the news of the tank reached the United States, people eagerly awaited details about the secret weapon, including its origins. Then, three days after the first tank action in France, Murray M. Baker, vice-president of the Holt Manufacturing Company in Peoria, fired a journalistic shot heard around the world. In an interview in Washington, D.C., he told reporters that it was Holt machines weighing 18,000 pounds and putting out 120 horsepower that had hurdled German trenches and faced intense enemy gunfire. Continuing, Baker reported,

> We have sold about 1,000 caterpillar tractors to the British government, but we had nothing to do with putting armor on them or placing machine guns, but some of our men at Aldershot, England, recently were notified that the British government intended to armor some of the tractors and use them for works other than the usual towing of guns. (25)

While Baker's statements to the press were reasonably honest, they did not make a clear distinction between the Holt crawler tractor and the British-made crawler tank. His interview left the impression that they were one and the same thing, and this is what newsmen conveyed to the public. Possibly because of wartime security

Our artist's conception of the armored "Armadillos" in action on the Somme. Based on reports of eyewitnesses at the front

Armored Tractors in the European War

A New Weapon Which May Break up the Deadlock of Trench Warfare

On October 12,1916 a public demonstration and parade greeted Ben Holt on his return to Stockton from a business trip to New York. Wide publicity on the use of Caterpillar crawlers in the Allied war efforts in Europe made Ben Holt a famous man. The belief by many that he had also invented the military tank added to his fame.

Artist's concept of the allied military tank after the Battle of the Somme in September 1916. The October 7, 1916 "Scientific American" article left the impression that the British tanks were adaptations of the Holt Caterpillar crawler.

The American press in 1916 presented artist's conception of the new military tank in Europe. The artist's imagination took various forms, often bizarre. Drawings were sometimes called photographs and the distortion became part of a propaganda campaign. The "Boston Post" on September 29, 1916 carried this version of the British tank in the Battle of the Somme.

This photograph in the "Chicago Herald" on October 19, 1916 was genuine and revealed the design of the first British tanks used on the Western Front in the Battle of the Somme in September of 1916.

measures, Baker may not have known that the British tanks had no American parts on them.

On September 19, the "New York Times" flashed the banner, "Famous Tanks of Somme Battle Had Their Origin at Peoria, Ill.," and the Stockton "Daily Evening Record" proudly exclaimed, "British Tanks Which Originated in Stockton Are Bringing this City World-wide Publicity." Most farm journals reported that the tanks were simply Caterpillar crawlers covered with heavy armor and machine guns.

Chief beneficiary of this avalanche of praise and publicity was Benjamin Holt, whose name appeared in most of the newspapers and magazines in the country as the father of the new invention. When it was later reported that it was the British who had actually modified the track-type principle to produce the tank, some chose to disbelieve the news while others emphasized that the new weapon was the original Stockton Caterpillar track-type tractor with only a few embellishments. (26)

Thrilled to have a true hero in their midst, the people of Stockton prepared to honor their adopted son upon his return on October 12, 1916, from a three-month business and pleasure trip to the East Coast. Plans for the surprise reception were made by officials of the Holt Manufacturing Company, the Rotary Club, the Chamber of Commerce, and other organizations, who enlisted C. Parker, Charles Neumiller, and Dan Gilmore to meet the Holt train in Sacramento. While driving into Stockton, talk was turned to a new rice harvester just being completed, and Ben of course said, "Take me to the factory before I go home. I want to see that harvester." (27)

This is what everyone had hoped for. As they reached the factory, a huge banner over Aurora Street read, "Columbus discovered America, but Benjamin Holt invented the Caterpillar." Over 2,000 Holt workers and townspeople gathered at the factory site to honor Stockton's leading citizen. A procession from the factory to the Holt residence was led by the Caterpillar band, followed by Benjamin Holt in a decorated automobile, and a five-block procession of people carrying red Japanese lanterns. It was one of Stockton's greatest public celebrations.

As a result of the nationwide recognition afforded Benjamin Holt and Caterpillar, the Holt Company found itself swamped with orders and sales, and the name Caterpillar became familiar to the entire country. Holt business net earnings rose over 100 percent from $561,000 in 1915 to $1,136,000 in 1916. So widely known was the Holt Company that a personal representative of Czar Nicholas II traveled all the way to Stockton to place orders for farm machinery. Rumors spread that he had already purchased 1,000 big track-type tractors at a cost of $2,500,000, but this story was denied by company officials. (28)

At the company's annual meeting in 1916, Holt directors reported the most profitable year in their history, predicting that business would double in 1917, and declared an annual dividend of 7 percent on preferred stock and 10 percent on common stock. The payroll now numbered 2,100 employees, and a new foundry, motor assembly, and harvester assembly shop had just been completed. (29)

Flushed with these good times, the company gave their Stockton employees a Christmas bonus of gold coins from the San Francisco mint. Just around the corner, however, was the end of 29 months of neutrality for the United States in the European war.

Chapter 15

THE WAR COMES TO STOCKTON AND PEORIA

When Congress passed the joint resolution declaring war on the German Empire on April 6, 1914, the American people pledged their blood, treasure, and sacred honor, to a crusade to make the world safe for democracy. Although the battleground was in Europe, the war was soon to become part of the lives of the people of Stockton and every family in the United States.

When German submarines threatened food supplies going to the Allies, factory workers were urged to raise vegetables in back yards and vacant lots. They were informed that the Americans wasted a billion dollars of food annually by improper cooking, waste, spoilage and rodents, and that accordingly, potato peelings should be saved and used cooking grease turned over to the government for making munitions. While Food Administrator Herbert Hoover called on farmers to increase production, reduce waste, and keep pigs to empty the slop bucket, newspapers in Stockton urged people to "Hooverize" " by observing wheatless, meatless, and heatless days of the week. One newspaper cartoon showing a farmer with his hands on the plow was captioned "The Greatest Trench Digger in the World," while another depicted an American flag-wrapped girl telling an elderly man, "You may be too old to fight, but you're not too old to hoe." (1)

Major General Ernest Swinton of the British Armed Forces salutes Benjamin Holt on the factory grounds in Stockton, California, on April 18, 1918. A mock baby tank had been built for the occasion as a symbol of the new military weapons used in the war. It was powered by a motorcycle engine. To the crowd of 2,500 employees watching, the general made a striking figure in his army uniform, leather puttees, gloves and walking cane. Benjamin Holt wearing a dark suit, small bow tie, gold watch chain, and a soft felt hat looked strong and serious-minded.

Before a large crowd in Stockton, April 1918, General Swinton thanked Benjamin Holt and his associates for building the Caterpillar track-type tractor because the use of one in Belgium in 1914 gave him the basic idea for building a military tank used in trench warfare on the Western Front in Europe.

After the United States entered the war in 1917, the American people were asked to buy Liberty Bonds to help finance the Allied cause. In Stockton, the names of those who responded were published in the local newspaper, producing lists filling three full pages. In one early Liberty Bond drive, Benjamin Holt purchased securities totaling $25,000. (2)

Bond drives utilized every conceivable advertising technique to encourage sales. A typical full-page ad in the Stockton paper showed an avalanche of silver dollars crushing a prostrate Kaiser. The cost of this advertising space was donated by the banks of Stockton, the Holt Manufacturing Company, and the Sperry Flour Company. (3)

Intense patriotic fever was generated and during a mammoth parade of Stockton school children in support of the third Liberty Bond drive on April 8, 1918, wildly cheering townspeople turned out to watch the spectacle. By one newspaper account, "There were bands blaring, drums throbbing, fifes screaming, and the sturdy tramp of thousands of feet and cheers until the heavens echoed back the glad cry and passed it far across the seas to stir the yeast of democracy working in the ranks of a startled and blood-blinded autocracy."(4)

Because the Holt plant in Stockton manufactured tractors used in the war, special security was provided for the factory, especially after rumors circulated that enemy spies were busy sabotaging railroad bridges, munition plants, and factories. As a result, 18 soldiers from the San Francisco Presidio arrived in Stockton on April 20, 1917, to patrol the area around the Holt Manufacturing Company. When these patrols were withdrawn in February, 1918, the company built a 12-foot board fence topped with five strands of barbed wire to surround the plant. This mile-long stockade was erected in one day by 200 factory workers using 100,000 feet of lumber. Given a free dinner for their assistance, one worker remarked, "It wasn't the dinner we cared about; it was the spirit of the thing—something patriotic you know."(5)

Additional security was achieved at the plant by requiring every person who entered the factory grounds to wear an identificaiton badge. These badges featured pictures of the employee and Charlie Starkey, an old factory hand who had been with the firm since 1884. The double photograph badge was conceived to foil any attempt to manufacture a fraudulent pass. A diligent security guard once refused entrance to Benjamin Holt when he forgot his badge. (6)

The company's patriotic spirit found expression in a huge new flag which hung from a 100-foot steel mast, and workers in individual departments of the factory donated money for the purchase of flags to fly above their own buildings. One worker recalled that the machine shop first bought a large flag, then the mill block followed with a larger one, and finally the workers in the main office bought a flag so large that it required a

The Holt assembly line in the machine shop during the war years.

double-jointed, reinforced pole to hold it aloft. In addition, a large service flage that carried 280 gold stars, was hung over Aurora Street representing the Holt workers who had gone to war. (7)

Patriotic feelings were regularly kindled at noontime rallies after the big steam whistle above the engine room blew, which under favorable weather conditions could be heard a distance of 15 miles. During the noon hour the Holt Band played martial music, and special speakers addressed the workers about the importance of cooperation in the war effort. Company lawyer Charles Neumiller often spoke to the crowd and, like a football coach at half-time, would try to arouse the workers to greater production efforts on behalf of the war. (8)

"Four Minute Men" sometimes spoke at the noontime rallies. These volunteer local speakers—75,000 across the United States—had been trained by the National Committee of Public Information to deliver fiery four-minute speeches to factory, civic, and church groups. Their pre-written speeches were designed to whip up support for the Red Cross, bond drives, and various other relief agencies. (9)

The Holt Company band in Stockton frequently "struck up the band," during the war years.

Recognizing that their employees were working at maximum effort, Holt Company officials, with the sanction of the government in Washington, D.C., continued to hold the traditional, annual company picnic as an extra holiday throughout the war years. The 1917 picnic began with a morning parade of the entire work force through the streets of the city to a park where baseball, boxing, wrestling, foot races, a tug-of-war, nail-driving contests, and a greased pig chase were scheduled. Merry-go-round rides, a band concert, speeches, and dancing went on through the afternoon and evening. A photographer took moving pictures which were later shown in a Stockton movie theater. Although families brought their own lunches, the Holt Company furnished ice cream, soda pop, peanuts, pop corn and coffee to the 9,000 people in attendance. Newspapers unselfconsciously noted that on this festive day, "colored employees and their families left on a special train in the morning and reached Salomon Grove on the Tidewater Southern route where a very enjoyable day was spent." (10)

Throughout the war, the 2,100 Stockton Holt employees worked a six-day, 48-hour week with some additional overtime and Sunday shifts. Coffee breaks, paid vacations,and other fringe benefits were unknown. In 1917, unskilled workers received 35 cents an hour, and machinists operating Lodge and Shiply lathes and Milwaukee milling machines earned 50 cents an hour.

Working conditions in the Holt shops had improved, until by 1918 there were two nurses on duty to care for minor injuries and a doctor who held office hours during the week. Under the new worker compensation laws in California, the company paid the cost of medical care caused by injuries on the job. (11)

In the winter, the harvester shop was heated by burning charcoal in oil drums, but most of the buildings in the five-block area received heat from noisy steam pipes. A few buildings had no heat at all, which according to one employee, was why Ben Holt wore his old overcoat so much of the time. A company-owned cafeteria provided meals, and although there was a cigar store across the street, Benjamin forbade smoking in most areas of the factory because of the fire hazard. (12)

The Stockton plant was almost self-sufficient. It had its own printing department, electrical shops, carpenters, plumbers, painters, musicians, millwrights, and volunteer fire fighters. On one occasion a fire in the motor-testing room threatened to destroy part of the factory, but trained volunteers successfully controlled the blaze until the Stockton firemen arrived. The company sent the city firemen $500 in appreciation of their quick service. (13)

While most of the Stockton employees approved of the company's labor policies, a few malcontents decried the management's paternal treatment of its workers. They also objected to the company's banning of unions at the Stockton factory while permitting them in the newer Peoria plant. In addition, a few pacifists critized the Holt Company's contracts with the War Department. In an editorial in the "Stockton Labor Review" on June 13, 1918, Dilse Hopkins described the workers in the Stockton plant as brow-beaten slaves marked with identification photographs and numbers on their chests like convicts in prison and forced to march in city parades and attend company picnics to retain their jobs. Outraged citizens and government officials immediately

Assembling tracks at Holt's East Peoria plant in Peoria, Illinois about 1918. All the motors were manufactured in Stockton and shipped to Peoria for final assembly. United States Government inspectors checked production results.

During World War I, Pliny Holt worked as a "dollar-a-year man" for the United States Army Ordnance Department on the design of self-propelled mounts for military weapons such as howitzers and light field guns. This picture shows a track-type tractor carrying a 75 millimeter Mark VI gun. The machine was demonstrated in San Francisco in 1919.

demanded a retraction of this editorial. This was forthcoming within the week when the editor probably realized that he might be sent to jail under the provisions of the new federal Sedition Law approved the previous month. (14)

In Stockton, the wave of 100 percent Americanism, fostered in part by government propaganda, resulted in the denial of citizenship to a Hindu business man who had lived in the town for 15 years, and letters to the editors from farmers who preferred to see their fruit rot than have it picked by Oriental workers. The "Stockton Evening Record" itself urged readers to shun conscientious objectors because they "had long, lean, sallow faces."(15)

Some officials and employees of the Holt Manufacturing Company were especially caught up in this patriotic crusade, and Charles Neumiller, for example, became a leader of the local chapter of the American Defense Society. This organization supported universal military training, enforcement of the Monroe Doctrine, the death penalty for traitors, exposure of German atrocities, and total respect for the nation's flag. While most of the goals of this society were meritorious some of their methods were extralegal.

When this enthusiastic group met in the Stockton Auditorium on the night of March 29, 1918, a crowd of 1,000 people was told that any man who refused to stand and remove his hat when the "Star Spangled Banner" was played should be hit in the nose. Neumiller exhorted that "the man who mutters against the flag, whether in the home or in the street, must be suppressed now and forever," and a local judge urged that everyone consider himself a committee of one to ferret out disloyal citizens and German sympathizers. To extended cheers, the judge further endorsed the lynching of German spies.

After the mob had quieted down, a soloist sang "Keep the Home Fires Burning" and the French Naitonal Anthem, while a journalist observed that probably never before in the history of the city had the stirring songs been listened to with such devotion. Another speaker advocated that all members of the Industrial Workers of the World be banned in Stockton and their building moved out of town, a vigilante action wildly endorsed by the crowd. (16)

For most Stockton residents, however, life continued almost as normal during the war years. Since gasoline was not rationed, people took their automobiles on sightseeing trips, to visit friends, or to go fishing. Baseball teams in Stockton played a full schedule.

Production of war materials nevertheless remained paramount., After the declaration of war, the reorganized Holt Company placed all its facilities at the disposal of the Federal Government and followed the instructions of the War Industries Board directed by Bernard Baruch. Little conversion of its factories was needed, because it had been selling tractors to France since 1914, and to Russia and Britain since 1916, and track-type tractors to the Ordnance Division of the

United States Army prior to America's entry into the war. (17)

During the course of the war, the Holt Manufacturing Company signed a total of 66 contracts with the United States Government for the production of materials essential to the military conflict. These agreements ranged from the publication of instruction manuals for operating tractors in the war zone to contracts for producing 1,800 10-ton tractors. The first major contract signed on July 31, 1917, included an order for 90 120-horsepower track-type tractors, and 1,500 55-horsepower crawler tractors. Another contract authorized the use of the Holt patents in the manufacture of tractors in the United States Army arsenals. Although the amounts of money involved in these contracts were seldom disclosed to the public, the Peoria "Star" did note on one occasion that the Holt Company had received a $12,000,000 government contract for tractors. (18)

Unquestionably, the Holt Company held advantages over other tractor firms in securing Government contracts because of prior involvement with ordnance officials in the War Department. Holt executives had urged Army officials to test Caterpillar crawlers as early as 1913, and Pliny E. Holt had consulted with Major James B. Dillard about military ordnance in October of 1916. M. A. Chaplin, an engineer with the Stockton plant for many years, had supervised the operation of the Caterpillar track-type tractors on the Pershing Expedition into Mexico in 1916 and later became a major in the United States Army. (19)

In fact, the Holt organization had such a head start on all other manufacturers that when the Army Ordnance Department on February 8, 1916, asked for bids on five gasoline tractors, the specifications called for tractors with four cylinders of 45 horsepower "of the Caterpillar type." Subsequent specifications not only insisted that the tractors be of the crawler type but that they be "the Holt 45-horsepower Caterpillar or its equal." Even before the United States entered the war, then, the War Department had adopted the Holt tractor as the authorized standard model. According to Caterpillar's Murray Baker, when bids were submitted on Government contracts, former Holt employees were in the inner offices making decisions while other companies' agents were sitting in the outer offices cooling their heels. (20)

The Government's burgeoning war production programs obviously required considerable expansion of the nation's factory facilities. Active in war production work, the eight-year-old Peoria operation added a new machine shop in late 1916, celebrated by a huge factory banquet attended by Pliny E. Holt and 3,000 employees. "The Peoria Star" described the scene orchestrated by Murray Baker: 155 girls from the Methodist Church served 40 hams, 750 chickens, 6 tubs of pork and beans, 240 gallons of coffee, 4,000 rolls, and 30 gallons of pickles. The Caterpillar Band played, flags were unfurled, and the audience sang, "If You Don't Like Your Uncle Sammy, Go Back Across the Sea." Speakers told how the Caterpillar crawlers had performed brilliantly on the Siberian steppes, the Brazilian pampas, the Dakota wheat fields, and in Northern logging camps and Southern sugar fields—and that now they were thundering over trenches and through forests, complementing the work of submarines on the high seas. (21)

Further expansion of the Peoria factory occurred in July 1917, when the buildings known as the War Plant were constructed, and still another complex of buildings, including a machine shop measuring 40 by 435 feet, was erected in September 1918. During this same period, $203,747 was spent on new buildings at the Stockton factory. The total expenditure at both factories during 1917 and 1918 for land, buildings, machinery, and tools amounted to $2,500,000. (22)

Although the Holt companies were adequately compensated by their wartime contracts, the business of simultaneously manufacturing tractors for Great Britain, France, Russia, and the United States became very complicated. Overlapping directives from federal officials, Army officers, and company executives created a bureaucratic snake pit. The Quartermaster's Corps distributed instructions for the building of a military vehicle, for example, which included 299 regulations and 300 specifications. (23)

Throughout the war, the Stockton plant concentrated on the manufacture of tractor motors, which were shipped to Peoria for final assembly before being moved to embarkation points along the Atlantic coast. Army officers were stationed at the Holt factories to enforce Government directives, and sometimes company officials had their way, while other times the government personnel had the last word. The major problem facing both parties was that neither had ever produced quantities of war materials in such a short time, and conversion to mass production of heavy machinery required huge quantities of raw materials, machine tools, skilled workmen and new assembly lines.

A report by H. W. Alden, lieutenant colonel of the Tank, Tractor and Trailer Division of the Ordnance Department, summarized problems as seen from the government's perspective. Initially, the report claimed, the Holt Manufacturing Company put forth more effort to fill the British and French government contracts than the ones with the United States War Department. Objections to this preferential treatment to foreign governments were raised, and after a conference between Government and Holt officials in Chicago on January 4, 1918, Holt agreed to maintain a certain ratio between foreign and domestic manufacturers until the foreign contracts were completed. (24)

Other problems, according to the Alden report, were a shortage of materials because of lack of follow-up on deliveries of raw materials, inept production control methods within the plant, workers with low morale, innaccurate blueprints, and a "lack of earnestness" on the part of company executives in getting materials into production. After July 6, 1918, a new factory manager succeeded in maintaining greater production, and subcontractors met their commitments in better fashion.

On the other hand, the Holt executives thought the Army officials lacked good judgment. They demanded 100 percent perfection in parts, made many unnecessary changes in design, and were obsessed with red tape. When they insisted that several lines of production be carried out at the same time, this fouled up production.

On one occasion, for example, the Army decided that a tractor's flat fan belt should be changed to a V-belt; after this order had gone down the line, a new directive stated the V-belt should be discarded and replaced with a gear-drive system. A final directive ordered the factories to return to the flat fan belts, accomplishing nothing other than delaying production for four weeks. (25)

An outbreak of Spanish influenza in 1918 also created production problems in war factories. Dan Gilmore, manager of the Stockton factory, wrote Pliny E. Holt on November 14, 1918: "We are still fighting the influenza. At one time we had as many as 410 men absent from the shop. Of course, in normal times we figure a daily average of 100 absentees."(26)

In spite of the obstacles, the Holt Manufacturing Company's production achievements during the war were impressive. For war uses, it produced and sold 2,981 tractors for the United States, 1,651 for Great Britain, 370 for France, and 80 for Russia, for a total of 5,082 tractors. Most were 15-ton and 10-ton machines. (27)

The Holt family was as thoroughly caught up in the war effort as the Holt Company. Benjamin's son, William K. Holt, was commissioned at first lieutenant in the United States Army and sent to France. On short notice, he was called upon to set up a school to train drivers to operate tractors pulling French 155 millimeter guns weighing 14 tons and mounted on four wheels. William lectured to the trainees about the handling of the machines, which were sometimes tricky to operate. He taught them to avoid jackknifing while going downhill, how to start the engines by opening the petcocks on each cylinder to reduce the amount of compression and thereby ease the cranking, and to trip the impulse magneto before cranking to avoid an explosive backfire that could injure the driver. After the Armistice, William K. was transferred to the Presidio at San Francisco in March of 1919, where the 10-ton Caterpillar crawlers were used in training. (28)

With the outbreak of war, Benjamin's nephew Pliny went to Washington, D.C., where he worked in the United States Ordnance Department on the design and development of self-propelled mounts for high caliber guns, howitzers, and light field guns. These self-propelled track-type gun mounts were first designed by Holt engineers in 1916, and some of them were delivered to the Army proving grounds even before the United States entered the war. In addition, Pliny and his co-designer, James Dillard, developed a self-contained Caterpillar unit with a gun mounted directly on the tractor chassis. The first vehicle of this type was the Caterpillar Mark I gun mount, which carried an 8-inch howitzer, weighed 58,000 pounds, and achieved road speeds of four miles an hour. No one knew whether a heavy gun could be fired successfully from a track-type vehicle, and one observer recalled: "What a thrill it was during the first test to see an 8-inch howitzer suddenly emerge from a dense thicket at the Aberdeen Proving Ground in the spring of 1918, come up to the line, fire a few rounds, and quickly disappear under cover . . . History in the making." (29)

In 1919, Pliny Holt carried on experimental work for the US Army. One gun mount carrying a 75 millimeter gun traveled at a record rate of 28 miles per hour.

Although the Armistice was signed about the time the first track-type gun mounts were completed, their success indicated that these machines would be the artillery of the future. When Pliny requested to return to Stockton, it was agreed that his army work should continue at the Holt Manufacturing Company. Here the Mark VII gun mount was built in 1919, a vehicle carrying a 75-millimeter gun that achieved the unheard-of speed of 28 miles an hour. These weapons were later equipped with a waterproof motor enabling them to cross rivers.

In January 1919, one of these waterproof gun mounts was driven from Stockton to San Francisco. It roared up Market Street and stopped in front of the Palace Hotel, where Mayor James Rolph gave a welcome and Mrs. Rolph smashed a bottle of champagne over the butt of the gun. After firing a blank shell in the direction of the Golden Gate, the gun proceeded to Ocean Beach, where it ran out through surf until it was submerged. As this "Racing Cannon," as it was called, roared about, an athlete challenged it to a three-block race, and lost the contest by 600 feet. Moving picture footage taken by major studios further publicized the military successes of Stockton's Holt Company. (30)

The war itself came to Benjamin Holt in Stockton when British Major General E. D. Swinton made a good will trip to the United States to elicit greater support for the Allied cause. Arriving in Stockton in April 1918, the general joined Benjamin on a platform constructed in front of the main office building where the company's 2,500 employees were gathered. Swinton acknowledged that he was pleased to meet Holt and be in Stockton, "the place where the Caterpillar Tractor was born." Benjamin Holt, who had been coerced to make this public appearance, replied, "General Swinton, you couldn't have come to a better country. I've traveled around considerably, and I've come to the conclusion that there's no finer spot. Welcome to California." (31)

Swinton's speech was full of praise for the Caterpillar crawler and the Stockton factory workers who had toiled and sweated to help the soldiers in Europe. Excerpts were circulated by the press and an expurgated version, minus the atrocity stories and rough language, was published in the May 1918 "Caterpillar Times". Photographers posed Benjamin Holt and General Swinton in front of a Caterpillar track-type tractor and a small, wooden mock-up tank powered by a motorcycle engine, which had been built for the occasion. The general made a striking figure in his army uniform, leather puttees, gloves, and walking cane. Benjamin Holt, wearing a dark suit, small bow tie, gold watch chain, and a soft felt hat, looked strong and serious-minded. Now 69 years old, his hair and well-trimmed mustache were white and his hands gnarled like those of a blacksmith.

When World War I came to an end on November 11, 1918, the struggle had taken the lives of 125,000 Americans and cost the United States approximately $26 billion. As news of the Armistice arrived, the factory in Peoria was closed for the day, and the employees given a full day's wages. According to factory manager Dan Gilmore, the Stockton factory went wild about the Armistice. When a premature announcement of peace had occurred on November 7, the workers had been given a half-day's vacation, but when the genuine Armistice was declared, thirty men and women from the factory core room left the factory without permission and marched down Main Street in their work clothes carrying a hastily made banner. Some workers quit their jobs on the spot, wrote Gilmore to Pliny, giving the impression that they had taken their war industry jobs only to avoid the draft. In conclusion, Dan wrote that he hoped Pliny would return to Stockton because there would be plenty to do in order to convert their two large factories to peacetime production. Only the Caterpillar "75" tractor was scheduled for continued production, a somber note for the company's future. (32)

Undisputedly, the war years had been profitable for the Holt Manufacturing Company. During 1917 the total assets of the company had increased $6.3 million, jumping another $6 million in 1918. The company's gross earnings in 1918 amounted to $7.4 million in domestic sales and $15.8 million in government contract sales for a total of $23.2 million in sales. (33)

During 1918, the Holt Company paid over $500,000 in dividends "after" 75 percent of the value of the Peoria plant had been amortized and more than $1 million in federal income taxes had been subtracted. In the same year, the company held $195,000 in Liberty Bonds, $720,000 cash in banks, and $6 million worth of inventory. From plows to swords and now back to plows, the Holt Company ended the war in an even stronger position than when it entered the "war to end all wars".

This 1920 photograph of a gun mount designed by Pliny Holt was probably taken at the experimental grounds of the Holt Manufacturing Company at Mormon Slough and Aurora Street, Stockton.

Chapter 16

POST-WAR ADJUSTMENT AND THE HOLT-BEST MERGER

Following the Armistice of November 1918, the American people looked forward to a period of peace and prosperity. This optimism was soon tempered by the realization that peace would be accompanied by some serious problems. Among them were inflation, which doubled the prices of consumer goods, and reconversion to a peacetime economy, which left three million workers unemployed.

Executives at the Holt Manufacturing Company had anticipated that adjustments to the post-war era would be difficult. Dan N. Gilmore, manager of the Stockton plant, observed immmediately after the Armistice that the company faced serious problems because it was operating two large factories with 5,000 employees but had only one profitable machine to offer the public, the 75-horsepower tractor. The company needed a smaller tractor to meet the demands of ordinary farmers. He wrote to Pliny Holt in Peoria, but the United States Army Ordnance Department had transferred Holt crankshaft dies for these smaller machines to a Midwest engine

This Holt advertisement appeared after World War I.

company, thus making it impossible to build large numbers of smaller tractors until they were returned. He added, "Something must be done immediately."(1)

The transition from war to peace was especially difficult for the Holt organization. Aside from the harvester shops, all the company's manufacturing facilities had been converted to making war machinery. During the last few months of the conflict, technical problems had finally been eliminated, and production was moving ahead at full capacity. In fact, Murray H. Baker had informed the War Department on July 26, 1918, that the contract for building 1,500 10-ton tractors would be completed in just five months because they were now rolling out of the Peoria factory at a rate of 400 per month.(2)

With the advent of peace in November 1918, federal officials announced that they planned to phase out all war contracts, a move which prompted some critics to complain that the "bone-heads in Washington" from the president on down were putting patriots in the position of suppliants. When Baker requested that all war contracts be honored, Colonel L. B. Moody of the Army Ordnance Department replied on June 23, 1919, that it was impossible to convince Congress that tractors should be manufactured for the Federal Government nine months after the Armistice, and Congress would in fact be cancelling all existing war contracts by June 30, 1919. (3)

As a result of this Congressional decision, six major contracts with the Holt corporation were terminated, including one for the manufacture of an additional 980 tractors and spare parts. Holt tractor sales under war contracts dropped from $10.9 million in 1918 to $5 million in 1919. The cancellation of war contracts drastically cut production at a time when the company had little commercial business to compensate for the loss. Between 1918 and 1922 the number of tractors manufactured annually in Peoria dropped from 2,580 to 868 and the number of motors built in Stockton dropped from 899 to 608. As a result, the number of employees at the Peoria factory was cut more than two-thirds from 2,307 to 707. (4)

Naturally this nose-dive in production was reflected in the company's total sales. In 1918, sales reached a high of $23 million; in 1919 they fell to $15 million, and by 1921 to $8.9 million. This represented a staggering decline of 66 percent in a span of only three years.(5) So substantial was this decline in business in the early post-war years that the company was finally forced to reorganize its entire structure and operations. In addition it even found itself with a new name and executive leadership.

While the Holt Company attempted to return after the Armistice to its original focus on manufacturing machinery for farm use, it was stymied in this effort by the nationwide financial depression lasting from 1920 to 1923, which hit farmers even more severely than other

This T 35 Caterpillar tractor is shown pulling two potato diggers on Schroder farm near Moorhead, Minnesota in the early 1920's.

sectors of the economy. When farmers suffered hard times, the Holt business also suffered. By the early 1920s, Secretary of Agriculture Henry C. Wallace estimated the farmer's purchasing power to be one-half of what it was before the war, and Congressman Robert LaFollette claimed that within a three-year period 600,000 farmers had gone bankrupt and farm values had declined $13 billion. One California farmer wrote a farm journal that he had been forced to work 16 hours a day to make a living, and despite these long hours he had "nothing to eat at our house except corn bread and bacon, until I have got bristles growing on my back."(6)

The Holt Company also suffered losses in sales of machinery to foreign countries during these years because of worldwide post-war economic depression. A sales manager of the Holt branch office in London reported in late 1922 that "business is bad and everyone is in a pessimistic mood." While the Holt corporation's export sales were 14 percent of its total sales in 1920, this dwindled to a nine percent in 1922.(7)

Fundamentally, American farmers and the Holt Company were experiencing the effects of the same problem—overproduction. Just as surplus crops glutted the market for farm commodities, so the wartime effort to produce more tractors finally backfired and flooded the power machinery market.

A 1920s photograph showing a Holt 5-ton tractor pulling three Maney scrapers across Grant Park in Chicago.

These Holt combined harvesters parked near a railroad in Stockton in 1924, were shipped to all grain-growing states in the US and to many foreign countries.

The Allied governments had contributed to this oversupply of farm tractors by encouraging increased production during the war years. In fact, the Ford Company's Fordson tractor was born out of World War I, when government officials in Great Britain and in the United States believed that more tractors were necessary to provide food to win the war. Over 6,000 new Fordson tractors were shipped to England in 1918, while Henry Ford boasted that he would soon build a million of them. He also said he would sell a Ford tractor, truck, and car together for less than $600, "if I don't croak first." Although he never kept this promise, he did add 650,000 tractors to the market by 1927.(8)

The market glut of tractors began immediately after the war when the United States government discovered that it owned 9,000 tractors, one-third of them Holt-produced machines. Fearing that these tractors would be dumped on the domestic market as surplus far below normal prices, Murray Baker suggested to Colonel L.B. Moody of the Army Ordnance Department on December 31, 1919, that these Holt tractors be turned over to state agricultural colleges for use in training farm operators.(9)

This idea was never acted upon, although the federal government did protect the market somewhat by distributing 1,410 of its tractors to the United States Engineering Corps, another 1,503 to army training camps, and a few to the Veteran's Bureau. Approximately 6,600 of the machines were then turned over to the Department of Agriculture which declared them "war surplus" and sold them to state highway departments for only a fraction of their original cost. Spare parts and machine tools were similarly sold for a song, with one set of dies originally costing $8,000 going to a buyer for $168. This "war surplus" action hit the Holt Company hard because the crawler tractors it was manufacturing were particularly well suited for road-building work.(10)

Even more disastrous for the Holt business was the fierce competition it experienced after the war from some new tractor companies which had been established between 1915 and 1920 to cash in on the profits anticipated to result from the rapid mechanization of agriculture. While some of the newly produced tractors had merit, many were freaks designed without working knowledge of farming. Three-cylinder motors, four-wheel drives, and wide drums instead of rear wheels were commonplace, in part because of the need to avoid design features already patented by the older companies like Holt.

Gambling that farmers would have to be weaned away from driving horses, some engineers built tractors which were driven by reins held by the farmer sitting on the plow or on top of his wagon load of hay.(11)

This motley array of new tractor models reduced the market for Holt track-type machines and confused farmers who were bewildered by the number of designs available. Most rural people were unable to distinguish between good tractors and junk— until they got home, and then it was too late. Not until the University of Nebraska inaugurated a tractor testing program in 1919 were buyers assured of any scientific data on which to base their decisions to purchase.

Despite this chaotic tractor market, the Holt organization remained solvent while hundreds of their rivals went bankrupt. This survival was more than good luck. The Holts were fortunate to build a reliable track-type tractor before competitors entered this type of production. In addition, the rigorous work on the Los Angeles Aqueduct and the action on the battle fields of Europe provided tests which enabled engineers to improve their machines.(12) Benjamin Holt usually stressed quality rather than quantity and he made repeated requests to his staff to machine parts with stronger construction. Furthermore he promised customers that they could always secure repair parts at reasonable of even original prices, a pledge which enhanced customer loyalty.(13)

During the 1920s, the Caterpillar Tractor Co. faced competition from other track-type tractor companies. officials in Peoria estimated that the number of tractors sold by competitors were as follows:

Apparently, Benjamin Holt was able to instill a sense of loyalty to the company by his employees. Most workers stayed on the job until illness of retirement forced them to leave. Throughout his 51-year career, he had shown management expertise as well as engineering genius.

The end of his life came suddenly and almost without warning. While living in Stockton, Ben had worked hard and carried heavy reponsibilities but he seldom drove himself to the point of exhaustion and he avoided excesses in his personal habits. The variety in his daily work gave him a change of pace, which, no doubt, was good for his health. Although he had experienced a few days of illness on occasion, he had encountered nothing serious and in 1920 he was expected to stay on the job. At 71 years of age he was still strong, energetic and worked without wearing glasses.

However, in November, 1920, he had not been feeling well although he spent much of his time in his shop. He became ill and was taken to Saint Joseph's Home, where his condition was thought to be serious, but not critical. Then, on December 4, he became gravely ill and died on December 5 at 11:00 in the forenoon. He had been in bed only ten days. Shortly before his death he asked questions about his latest engineering project in his experimental shop. The funeral was held in the Central Methodist Church with all members of the family in attendance. Alfred returned from his work at the Peoria

From 1915 to 1919, the Yuba Manufacturing Company of Marysville and Benecia, California built a 20-35 horsepower crawler tractor with a design which included the application of ball bearings to the endless track. The company claimed that the tractor would coast on plowed ground on a three per cent grade. However, if excessive wear occurred, these steel balls would scatter across the field.

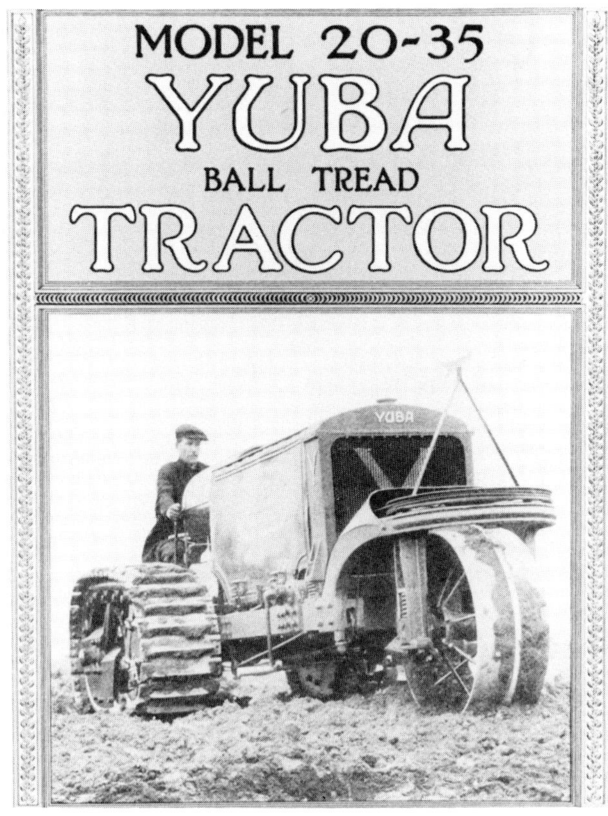

From 1910 to 1920 over 400 tractor companies entered the business of building farm tractors. Most failed because of faulty design, poor management and lack of capital. The Bean Trackpull tractor was built by the Bean Spray Pump Company, San Jose, California, in 1918. The tractor which was rated at 6-10 horsepower proved impractical.

plant, William K. came from Dallas, and the two younger sons, Dean and Edison returned from their studies at the University of California in Berkeley. Anne Holt Atherton, the daughter, lived in Stockton.

Meanwhile, A.C. Oullahan, the mayor of Stockton issued a proclamation directing that all business in the city be suspended during the funeral and that all flags be flown at half mast. In emotionally filled words he referred to Benjamin Holt as a great tree which had been cut down and thus this great landmark was gone. The people of Stockton should be loyal to his memory.

Benjamin Holt's death in 1920 knocked away the strongest underpinning of the Holt Manufacturing Company. Although executive duties had been shared by many, Uncle Ben had always served as the supreme court in making crucial management decisions because his extensive experience commanded wide respect among his co-workers and his judgments usually proved to be sound.

At the time of Benjamin's death, the House of Holt was prepared for his departure about as well as could be expected. His nephews, Pliny, C. Parker, and Ben C., had been with the organization for about 20 years. Uncle Ben's oldest son, Alfred, was in the business office in the Peoria plant, while his younger son, William K., was "learning the tractor business" at his father's request in Peoria.(15)

After the death of Benjamin, Thomas A. Baxter, a former Boston banker, was named president of the Stockton corporation. He had joined the firm in 1913 as a business manager. His role in company affairs became controversial as the next few years would demonstrate.

Baxter's ascendency in the Holt Company reflected the common corporate fact of life that even major corporations, however well established and well known, operate on large amounts of borrowed capital. Although the Holt organization included huge factories and annual sales amounting to millions of dollars, the firm's operations were dependent on large bank loans, or debt financing. Loans increased gradually from notes payable of $347,000 in 1911 to $3.7 million in 1917. In the latter year, the company was making monthly payments on loans of nearly $1 million from the San Francisco office of Bond and Goodwin, $2 million from the New York office of the same firm, and another $1 million from the eight California, Boston, Chicago, Spokane, and Stockton-based financial institutions.(16) By 1924, the standard pattern of borrowing heavily, paying off old loans, and re-borrowing had given the company some $5,675,000 in outstanding loans.

It is through this relentless need for expanding credit that financial institutions routinely became important in the internal as well as the external affairs of corporations. Financial advice leads to the securing of one or more seats on a company's board of directors, and finally (in some cases) an official from the dominant financial lender emerges as chief executive of the entire operation. This is what happened at the Holt Manufacturing Company.

Thomas F. Baxter was associated with the banking house of Bond & Goodwin, which had major offices in Boston, New York, Philadelphia, Chicago, and San Francisco. Baxter became acquainted with Benjamin Holt and his associates in 1910 when Bond & Goodwin made its first loans to the Holt Company. By 1913, Baxter's influence in the Stockton company had grown to the point that he was giving advice on such matters as the types of products to be discontinued, the system of bookkeeping to be followed, and the subsidiaries to be sold. In a 1913 letter to Ben Holt, Baxter delivered a classic banker's ultimatum regarding the ongoing need for financial capital:

> Under your agreement with Bond & Goodwin and the Wells-Fargo Nevada National Bank, a point will be reached somewhere in the next sixty days when further capital must be provided for your business, and unless prompt measures are adopted to bring about a combination and a reorganization on the lines above expressed, in my opinion, it will be impossible to secure the further capital required...otherwise, we cannot aid you in this respect.(17)

While no major reorganization was forced at this point, by 1915 the problems of managing the Holt corporation had increased so as to make major changes imperative. Pliny E. Holt, vice-president of the Stockton firm, wrote a long letter to Baxter on December 8, 1915, in which he expressed his concern about the need to enlarge the factory, to build an assembly plant in Canada, and to enlarge operations in Peoria. Expanded Peoria facilities would make it possible to reduce the excessive cost of freight on goods shipped east from Stockton, which amounted to $9 for every motor built in the Far West. Pliny continued that the company was losing a half-million dollars annually because of its inability to manufacture enough small tractors to meet the market demand. Confidentially he admitted that many compromises had been made by the board of directors to accommodate friends in salaried positions and that better leadership and clear lines of authority were needed in many areas. The company's experimental work should be separated from its manufacturing operations because "Uncle Ben simply will not make one machine at a time and follow it through to a successful finish...He jumps from one device to another and in that way never gets the benefit of the systematic development we must have."(18)

Accordingly, the need for a strong leader who would make tough decisions independent of individual family interests was acknowledged, and Baxter was hired as general manager of the Stockton factory in 1915. After Benjamin Holt's death in 1920, Baxter became president of the Stockton, Peoria based Holt Manufacturing Company, which he served until forced to resign in 1925.

As president, Baxter became a powerful figure disliked by most of the Holt family because of his authoritarian manner and hard-nosed business attitudes. He was elevated to the presidency after

Benjamin's death over the well-liked and knowledgeable Murray M. Baker because, according to Holt lawyer Charlie Neumiller, the amount of the company's outstanding loans to Bond & Goodwin left the board of directors no real choice. Baxter's ultimatum to the directors was like placing a 45 revolver on the table and quietly saying, "Gentlemen, do it my way, or there won't be a Holt Manufacturing Company."(19)

Baxter remained Holt Company president until 1925 when major stockholders, including Murray Baker, began secret negotiations with Harry H. Fair, an investment banker in San Francisco, to dump the Bond & Goodwin Firm. When Fair's group became more influential than the Bond & Goodwin faction among board members, the New York bankers lost their hold on the Holt Company and Baxter was unceremoniously ousted from control.

However unpopular Baxter may have been, he evidenced considerable success in bringing the company back into the domestic tractor market which had been ignored during the war years. The large 15-ton and 20-ton tractors that had been the firm's bread and butter during the World War I were phased out and replaced with the popular, smaller tractors designed and built between 1920 and 1924.(20)

Baxter recognized that the future prosperity of the Holt enterprises depended on sales large enough to keep the machinery humming on its 15 acres of floor space in the main factories in Stockton and Peoria. Aware that state and federal governments would be spending $1 billion on building highways in 1921 alone, he had the Holt advertising department begin tailoring its material to potential purchasers of road-building machinery. Large ads for earth-movers were placed in newspapers with large circulations, and additional advertisements were placed in road building, engineering, city administration, and other trade journals. About $110,000 annually was spent on advertising in the "Saturday Evening Post".(21)

Advertising copy in farm journals focused on the tractor's wartime service record in dramatic story form. Concluding messages noted that the "Caterpillar Tractors" with all the honor won on the battlefields of Europe, are at the disposal of American agriculture."(22)

Handicapped by slow sales in the post-war economic slump, enterprising Holt dealers attempted to capture attention with monumental publicity winning feats. On one occasion a dealer sponsored the climb of a 5-ton Holt to the snowy top of Pike's Peak, some 14,109 feet. The press, of course, exploited the feat just as the Holt dealer had hoped.

The Holt Company sales department was highly organized, with dealers and their salesmen given specific geographical regions. Dealers usually received a 5 percent commission on sales, while salesmen received a salary of $150 to $250 a month plus expenses.

Top saleman N. Gilmore, when working in Stockton, received a 10 percent commission on all sales in the eleven-county Sacramento Valley, over $2.3 million between 1918 and 1921, making his share more than $200,000. (23)

Because tractors, priced at $2,000 to $6,000, represented a sizeable investment even for a successful

The Tom Thumb Tractor Companay of Minneapolis, Minnesota built a track driven machine in 1915. The tractor had a Waukesha 4-cylinder motor, a Bennett carburetor, a dixie ignition system, Splotdorf spark plugs and a Perfex radiator and was rated at 12-20 horsepower. To make a turn the front part of the track was raised so the rear heel of the track could pivot on the ground. Its use was limited because of the odd mechanical design.

farmer, selling tractors was far from child's play. It took artful persuasion to convince a farmer, furthermore, that Holt crawlers were better than those made by C. L. Best Tractor Company in San Leandro, the Yuba Manufacturing Company in Marysville and Benicia, California, the Cleveland Tractor Co. in Ohio, the Bates Manufacturing Company in Joliet, Illinois, and the Bullock Tractor Company in Chicago.

Holt sales managers instructed their salesmen to tell farmers that Abe Lincoln freed the slaves and Ben Holt emancipated the horse. Salesmen were to be alert, neat, prompt, have a ready smile, avoid loafing, move quickly, speak in a low, distinct voice, and avoid gossip. They were never to mention hard times, and salesmen with a "yellow streak" who always complained were liable to be fired. One company memorandum stated, "If you want to make farm sales, hunt! Open gates, ring doorbells, sit on water troughs and fences, and wear out some chalk on the back of the tool shed."(24)

The sales departments' advice to salemen also included detailed information about the superiority of Holt tractors to other track-type machines. One set of salemen's instructions in 1923 contained 25 pages explaining how to combat each of 19 arguments used by the C. L. Best Company sales force to prove the merits of Best tractors. Points of comparison included radiators, clutches, transmissions, magnetoes, frames, steering systems, and motor bearings.

Testimonials were another much-used item in the Holt sales and advertising departments. In the early twenties, owners volunteered words of praise for the virtues of the Caterpillar crawler in preparing orchard land in the Rio Grande Valley of Texas, for farming blueberry swamps in Massachsetts, for grading streets around the country, for hauling in brick yards, for mining gypsite in Texas, for building railroads in Montana. Testimonials enthused about the Caterpillar crawler's suitability for nursery work in Sturgeon Bay, Wisconsin; for moving railroad locomotives in Pennsylvania; for building a golf course at Yale University; for hauling cinders in a glass factory in Hartford, Indiana; for farming muck soil in orange groves at Lake Alfred, Florida; for moving logs in Maine, Michigan and Idaho; for combing grain on the Palouse hills in Washington; and for tank-sinking, silt-scooping and wool-carting in Australia. (25)

Despite the fine performance of the Holt tractors, the competition generated by the nearby C. L. Best firm proved formidable. Disenchanted with his father's sale of the first Best factory to the Holt Company in 1908, C. L. Best resigned and set up another Best factory in Elmhurst where he produced his first track-type tractors in 1913. During the war years, the younger Best had concentrated on manufacturing farm tractors while the Holts were preoccupied with heavy-duty war tractors, and he thus was able to move ahead with little competition on the domestic scene. By the early 1920s, his Best "60" and Best "30" tractors had been refined until they compared favorably with the Holt tractors. Slightly more streamlined and carefully finished than the Holt machines, they still lacked Holt's worldwide reputation.

The Best Company compensated for this handicap by developing an aggressive sales force, with salesmen who gave liberal credit terms to prosecptive buyers and paid high prices for used tractors taken in trade. A Holt dealer in Oklahoma in 1924 complained that the Best Company was spending money like a drunken sailor and hiring three times as many salesmen and servicemen to assist its dealers. Another Holt dealer in Georgia claimed Best salesmen had offered to pay $1,000 for an old wheel tractor which was not worth $10 for junk. A St. Louis Holt dealer reported in 1922 that the state highway department had bought $73,000 worth of Best tractors in spite of his argument that the Holt machines were of

The largest four-wheel farm tractor on the market in 1916 was the Twin City "60", manufactured by the Minneapolis Steel and Machinery Company of Minneapolis, Minnesota. The 6-cylinder 60-90 horsepower tractor was advertised as the largest and most powerful tractor in the US. It weighed 27,700 pounds and had drive wheels 84 inches in diameter. 17-3

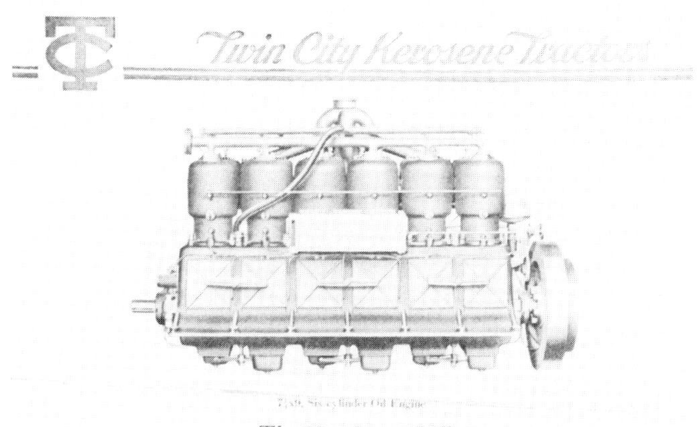

The Twin City "60"

THE Twin City "60" is the largest and most powerful tractor built any where in the United States today.

In its arrangement it is similar to the famous Twin City "40," the entire tractor, of course, being built more massively to be able to transmit the enormous power of its wonderful six-cylinder engine. This engine of 7¼" bore and 9" stroke is equipped with our Twin City vaporizing manifold, which enables it to operate most economically and with great flexibility on kerosene and other low-grade fuels.

It is guaranteed to develop over 90 brake horse-power continuously on the belt when

The smallest tractor manufactured in 1916 was the 5-10 horsepower Avery built in Peoria, Illinois. The 4-cylinder engine had a 4-inch bore and a 3-inch stroke and weighed 1,700 pounds. In 1916 60,000 people a day attended a demonstration of the tractor at the Fremont, Nebraska Tractor Show.

superior quality, while word came from Florida that Hardee County officials bought Best tractors because Best asked for only one-third of the sales price in cash, took promissory notes for an additional two years, and successfully bribed a county commissioner.(26)

Holt officials took the Best challenge seriously, noting that their rival had expanded from 15 dealers in 1919 to 50 in 1924 and the Best sales had risen 70 percent. In 1923, Murray Baker informed Holt Company president Baxter that the Best competition was the most difficult problem he faced. The Best "30", he conceded, was a good product, and because it was priced lower than the Holt 5-ton machine, it was one which would foster more competition.(27)

A Peoria sales report in 1924 concluded that 1925 would be "a most critical year. If the Best Company makes as much progress in the next 12 months as it has in the past year or two, our volume of business will be menaced..."(28)

Navigating in troubled waters, the Holt Company hired a consultant to make a comprehensive study of their entire industry and make recommendations for enhancing future business. The findings that went to Baxter's desk in October 1923 noted that although the company's 2,850 employees were manufacturing products which approximated $15 million in sales annually, the factories were operating at only 60 percent capacity because of insufficient sales. Secondly, the report recommended that fewer models of tractors should be built: the 10-ton should replace the 15-ton "75", a new 5-ton model should replace the old T-11 and the "T-35", should become the standard tractor. In addition, only two models of harvester should be produced. Thirdly, the report concluded that the business should be consolidated, the payroll reduced, and more interchangeable parts designed. Finally, more factory work should be done in Peoria because the freight rates on all material shipped to Stockton were a dollar higher per hundredweight. This cost exceeded the savings achieved in labor costs which were 10 to 15 percent lower in Stockton than in Illinois. This decisive 12-page report undoubtedly influenced the thinking of the Holt executives, who met on January 30, 1924 and drew up a set of resolutions regarding strict economies to be made in capital outlays.(29)

The heartbeat of the Holt corporation, of course, was still strong, with factories on 61 acres of land capable of producing 5,000 tractors annually. A company audit on July 1, 1924, showed that the firm had $12 million in assets and only $6 million in liabilities. Real estate, buildings, and machinery were worth $5 million, the inventory $.7 million, and cash in banks totaled $1.5 million.(30)

There were, however, two serious problems facing the management—the inability to secure enough sales to keep the factories operating at full capacity and the burden of carrying $5.6 million in short-term bank loans. Notes falling due in mid-1924 totaled $757,500, and over $2 million was payable to the banking house of Bond & Goodwin, Baxter's former employer.(31)

Financial affairs at the C. L. Best Tractor Co. in San Leandro were equally troublesome. With a productive capacity of 2,500 crawler tractors a year, its profits in 1924 were just under $1 million on sales totaling $5.3 million, almost half the total of $12 million made by the Holt Manufacturing Company in the same year.(32)

Like Holt, the best Company had borrowed heavily, primarily from San Francisco's Pierce, Fair and Company. The head of this firm, Harry H. Fair, had bought stock in the Best corporation and secured bank loans for it for many years. Concerned about the Best Company's ability to meet these bank loans, Fair entered into secret discussions with the Holt executives about a merger of the two corporations. On March 2, 1925, Murray Baker released a telegram to the press which contained the surprising news of a mammoth financial regrouping of the Holt Company with the Best Company through its principal shareholder, Pierce, Fair and Company:

> At a meeting of the Board of Directors of the Holt Manufacturing Company today, Thomas F. Baxter, for the past five years President of the Holt Manufacturing Company, resigned, and there was elected in his place Charles L. Neumiller. The control of the Holt Manufacturing Company, which for the last 30 years has rested in the Holt family and more recently in the Benjamin Holt Corporation, has passed to a group consisting of Pierce, Fair and Company of San Francisco, and associates. In this group are included practically all of the members of the Holt family, M.M. Baker of Peoria, Illinois, and the Benjamin Holt Corporation.(33)

According to the company vice president in Peoria, this new arrangement represented not a bona fide merger, but only a consolidation of the two machinery companies in which the buyers of the new stock would be restricted to those who already owned stock in one or the other corporation. Neither the Holts nor the Bests had sold out; both groups with interlocking interests had merely reorganized their management and financial ownership.(34)

After the agreement on basic principles of the merger, the Holt stockholders met to approve the sale of all the assets of the Holt Manufacturing Company to a new corporation called the Caterpillar Tractor Co. The owners of the Best Gas Tractor Company did the same at a similiar meeting in San Leandro.

The new corporation had assets of nearly $12 million and a floating debt of $4.2 million. Its total sales were expected to be $17.5 million in 1926, with earnings in excess of $2.2 million.

In the new company's executive hierarchy, Best people dominated the administration by holding two key positions: the chairman of the board and president of the company. Only vice-presidents Murray M. Baker and Pliny Holt represented the Holt faction on the board of directors, although C. Parker Holt was made manager of export sales after being elected to the board of directors in 1928. (35)

Among the commonly offered reasons for the enforced merger was the belief that Benjamin Holt had employed too many less-than-qualified relatives, nepotism which had been detrimental to the welfare of the company over many years. In May 1938, "Fortune Magazine" repeated the comment as if it were repeated the comment as if it were gospel.

While there may be truth in this generalization for the company's later years, the early success of the Holt enterprise was in fact based on the cooperation of the original Holt brothers and the three nephews, C. Parker, Ben C., and Pliny E. Pliny's contributions were primarily in engineering and war work, Ben C.'s services were managerial in the Pacific Northwest, and C. Parkers's contributions were in sales and as a businessman. Other family members and friends exhibited industry, skill, and dedication, with a cooperative spirit rare among the annals of business history. Nepotism may in fact have kept the company going rather than brought it to its knees.

While the Holt-Best merger in 1925 removed the Holt name from the corporation's masthead, the new larger corporation eliminated the intense competition between the two giant manufacturers of track-type tractors. It also ended protracted legal struggles over patent rights, enabling the new corporate entity to produce greater profits for its owners and stockholders. It was also a move toward concentration of resources seen in many other growing United States industries in the same period.

The soundness of the decision to merge was quickly affirmed by all economic indices. Total sales of the new Caterpillar Tractor Co. were $14 million in 1925, $21 million in 1926, and $52 million in 1929. Profits escalated from $4.3 million in 1926 to $12.4 million on the eve of the stockmarket crash in 1929. (36)

For the company's home city of Stockton, the merger had lasting consequences. Following consolidation of the company in 1925, the manufacture of all tractors was transferred to Peoria, and a subsidiary, the Western Harvester Co., was established in the old Stockton plant to continue building combined harvesters. With the advent of the Great Depression in the 1930s, this work was discontinued and all production shifted to Peoria.

Today the boarded-up, corrugated-iron buildings of Ben Holt's factory still fend off the winter rains in Stockton, marking the site of the city's contribution to the nation's technological renaissance. Benjamin Holt's original experimental shop has been removed to Stockton's Haggin Museum, where visitors can see the remains of the historic laboratory that brought machines to America's farm land.

Chapter 17

HOLT GOES TO COURT

Just as debt financing and business mergers became major facts of American corporate life in the twentieth century, so litigation emerged as another feature of modern business practice.

At the Holt Manufacturing Company, litigation focused on patent infringements. Protracted court battles over patents became part of doing business after 1900, and only those companies with a gladiator spirit survived the challenge of going to court or going out of business.

Benjamin Holt was no passive bystander in these pitched business confrontations. Instead, he exercised tough-minded leadership, defending his patent rights tenaciously against infringements by would-be challengers. Once provoked to action, he sunk his teeth like a bulldog and poured money into legal cases that often dragged on for years.

Inventor's patents, issued for a period of 17 years, were highly valued by manufacturers as a primary means of protecting their companies' economic investments. But patents were subject to frequent litigation because of the difficulty in drafting patents accurate and precise. As a result, patent litigation often rested on the interpretation of a judge and jury, thereby leaving corporations vulnerable to a host of lawsuits, some legitimate but many prompted by questionable motives.

It was the intense rivalry between Daniel Best and Benjamin Holt in building combined harvesters and steam traction engines in the 1890s that led to the first big lawsuit involving the Holt Company. In May 1904, Daniel Best filed a claim in the United States Circuit Court for the Northern District of California alleging that the Holt Manufacturing Company had infringed on his patent of 1889 for an arrangement on the combined harvester that used an auxiliary steam engine which received steam from the boiler of the main traction engine. Best asked for the payment of $100,000 in damages. (1)

After three years of pre-trial preparation and legal maneuvering, the case went to trial on November 21, 1907. It has historical interest because the witnesses included the Best's most knowledgeable experts in farm machinery. Prominent inventors such as Daniel Best, C. L. Best, Pliny E. Holt, and George S. Berry explained under oath the design and uses of their machines in great detail. (2)

Daniel Best had built his first steam traction engine in 1889. It developed 100 horsepower, and to prevent the harvester from choking down in heavy grain, he conceived "the idea of putting an auxiliary steam engine on the side of the harvester . . . I got the idea in 1887, but did not perfect it until 1889 when I got a patent on it . . ."(3) Best's invention was one of the first "power take-offs" in mechanized agriculture, an important technological advance. His machine carried steam from the main engine boiler to an auxiliary engine bolted on the frame of the combined harvester through a flexible pipe with a "ball coupling" to give it maximum flexibility.

In the ensuing trial over Best's patent, witnesses testified that the Holt Manufacturing Company had copied the Best patent of 1889 by building steam harvester outfits that featured an auxiliary steam engine bolted to the rear part of the engine frame. This engine carried the power through swivel-universal joints back to the threshing apparatus of the harvester. Holt witnesses said, however, that the Benjamin Holt patent of 1896 did not infringe on the earlier Best patent because the extra engine was never placed on the combined harvester. It was placed on a platform at the rear of the steam traction engine.

According to the plaintiff's witnesses, the Holt harvesters made from 1896 to 1904 were unsatisfactory because the universal-joint fixture frequently broke when the engine made sharp turns or traveled over rough ground. Therefore in 1904, they said, the Holt Company,

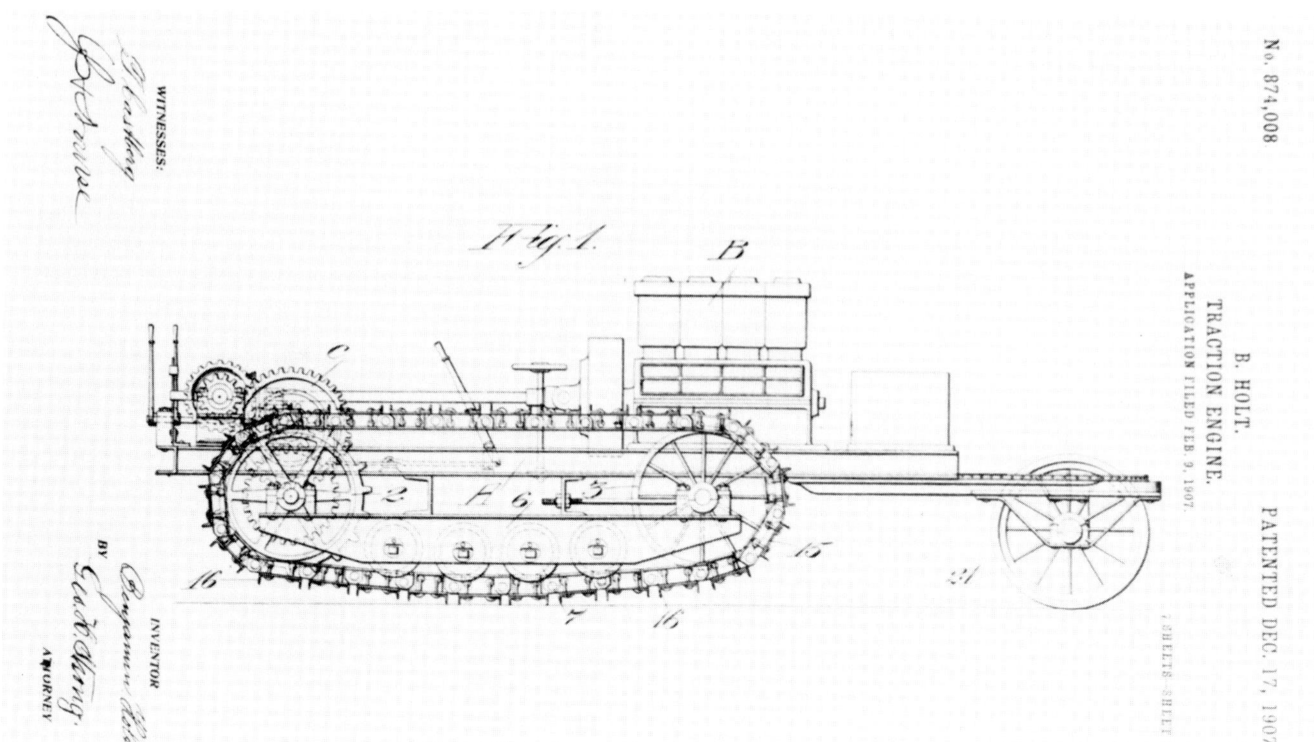

Drawing of a December 17, 1907 traction engine patent issued to Benjamin Holt.

like Best, placed the auxiliary engine on the front of the harvester frame and piped steam back to it. This infringed on the Best patent rights, and Best was entitled to damages.

Another witness, J.H. Davis, who had worked as a mechanic for Benjamin Holt for 13 years, testified that in the spring of 1901, the Holt Company began moving the auxiliary engine from the traction engine to the harvester. On one occasion, according to Davis, Benjamin Holt expressed qualms about the change because Daniel Best was building his machines this way. Finally, however, Holt exclaimed, "Damn it, we will put it on, and if we have to pay for it, we will pay for it. That is all there is to it." (4)

After two weeks of testimony in this case, much of it highly technical in nature, the jury brought in their verdict on December 19, 1907, assessing $35,000 in damages against the Holt Company. Apparently the Holt Company had not presented a strong case and had even failed to show that Daniel Best's patent of 1889 was not original. (5)

Stung by this adverse decision, Holt lawyer, I. M. Kalloch appealed the case in early 1908. He charged that 26 errors had been made during the trial and that in fact Daniel Best's patent of 1889 did not represent what is known as a "pioneer" patent. (6)

Kalloch's strongest evidence for the impropriety of the Best patent was the work of George S. Berry of Tulare County, California, who had patented a self-propelled steam combined harvester on December 6, 1887, which carried an auxiliary engine activated by steam drawn from the boiler of a traction engine. Five of the machines were used on his 20,000 acre ranch. (7)

Berry's harvesters included all the features of the Best machinery. They cut heavy grain, they turned square corners, they operated independently of the main traction engine, and the harvester could be detached from the outfit so the engine could be used for other purposes. There was nothing novel in the Best patent, claimed the Holt lawyer, other than the minor matter, certainly not a pioneer invention, of the location of the auxiliary engine in the center of the combined harvester. To claim this small feature as a pioneer invention was unsound. Furthermore, if George Berry's patent was valid, then no one else could obtain a similar patent until it had expired. Thus Daniel Best had merely improved on a prior art, and if Best could make such an improvement without penalty, then Benjamin Holt could do likewise. If Best had the right to use Berry's concept and place the extra engine in the center of a trailing harvester, then Holt also had a right to place his auxiliary engine on the front of the harvester.

After filing these documents with the United States Circuit Court of Appeals in May 1908, both sides waited eagerly for the decision. Finally in August 1909, Judge Gilbert Ross of the Circuit Court of Appeals reversed the verdict in the Best-Holt Case and remanded it to a lower court for a new trial. (8)

In the intervening years, however, the relations between Holt and Best had changed drastically. In 1908, while both sides were waiting for a decision on the writ of error, the Best Manufacturing Company was sold to the Holts for $325,000. (9)

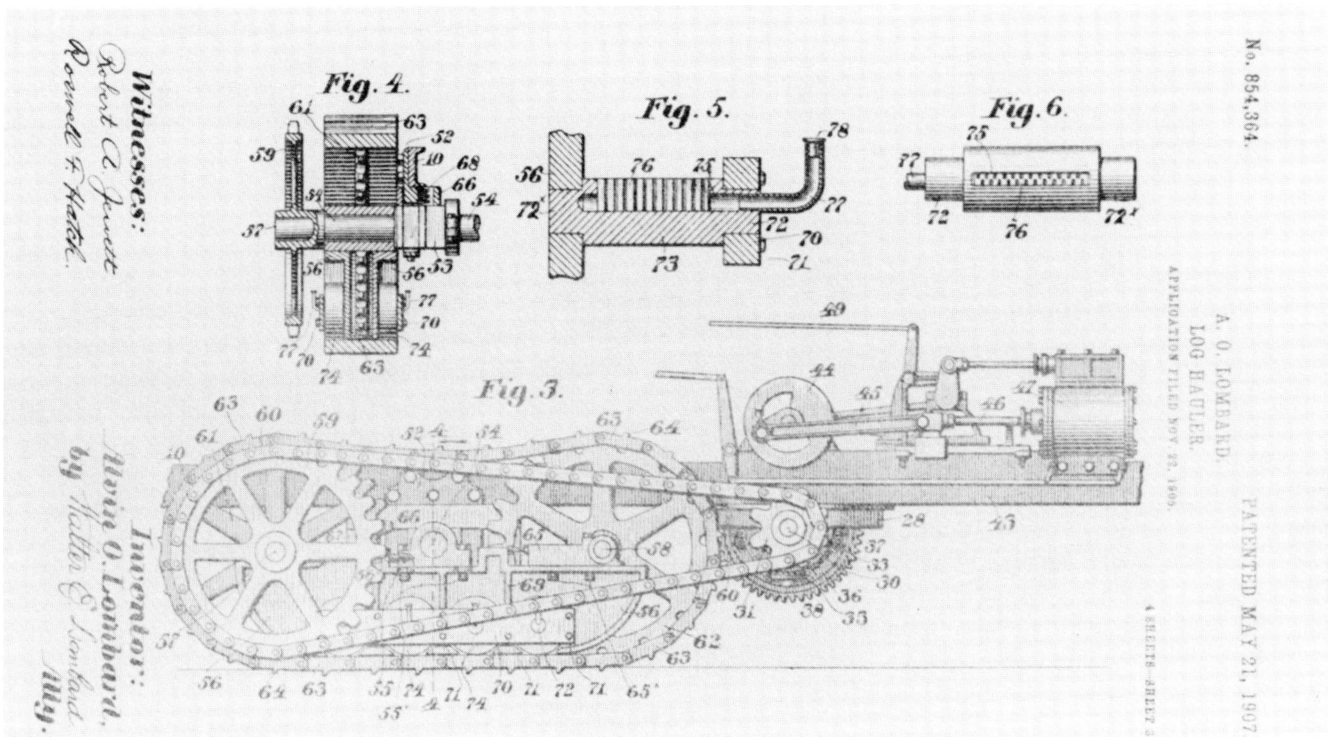

The Lombard log haulers were driven by chain drive systems. This Lombard patent was issued on May 21, 1907.

Since the Holts now owned the Best factory and since all parties were weary of the matter after five years of litigation, both sides agreed to drop the case, assume their own costs, and end the battle over auxiliary steam engines on combined harvesters. The old steam technology was being pushed off the scene anyway by the 1910s, and gasoline engines were moving to the fore as the primary source of mechanical power.

This unexpected lull in the legal wars between the Best and Holt families proved short lived. Like burning embers smothered for a moment, the Holt-Best competition soon flared up again. Having clashed over patent rights for combined harvesters, they now moved on to wrestle with similar controversies in the manufacture of track-type gasoline tractors. But now the stakes were higher.

The second major trial between the two antagonists occurred when Daniel Best's son resurrected another Best Manufacturing Company. The first Best Company had been sold to Holt in 1908 with the agreement that C. L. Best would then buy one-third of the stock and in return be made manager of the San Leandro plant. (10)

C. L. Best, however, refused to pay for his share of the stock and resigned from the Holt firm in March 1910. On resigning he agreed that he would not engage in manufacturing harvesters, traction engines, or freighting outfits for a period of ten years, but he in fact immediately incorporated his own new business, the C. L. Best Gas Traction Company, borrowed money from his father, and built a new factory in Elmhurst, one mile distant from the old Best plant in San Leandro. The friction between the two corporations continued to grow, because now both the Holts and C. L. Best were manufacturing crawler tractors.

In 1912 Best published a catalog which advertised a 30-horsepower crawler tractor which did not yet exist and contained illustrations of harvesting scenes in which half of the machines were Holts. In addition, the catalog used the word "Caterpillar," the registered trade name for the Holt tractor. (11)

In addition the Holts discovered that the new C. L. Best Tracklayer was a carbon copy of their Caterpillar model in mechanical design, function and appearance. The four independently-cast motor cylinders of the valve-in-head design developing 75 horsepower, the two tracks attached to a supplemental frame, the single forward steering wheel, the canopy top, the power steering, and the fuel tank mounted on the left side all reflected the Stockton predecessor.

When Best aggressively advertised his tractors at the Panama-Pacific International Exposition and the State Fair in 1915, Benjamin Holt advised his legal department to file a suit against the C. L. Best Gas Traction Company. The suit charged infringement of the Holt patent of December 17, 1907.

This action was not taken on the spur of the moment. Previously, Company officials had conducted ongoing studies of the patent history of track-laying vehicles. As early as August 1908, company lawyer Charles Neumiller sent copies of Alvin O. Lombard's two basic patents of 1901 and 1907 for lumbering tracklayers to patent attorneys Kalloch & Fox in San Francisco, asking whether the Benjamin Holt patent of 1907 (Number 874,008) infringed on the Lombard patents. These lawyers had replied that the Holt tractor did not infringe on any of the Lombard claims in his first patent of 1901 because the Lombard engine was carried upon a pair of roller-belts while the Holt engine moved on a number of small wheels which carried the weight of the machine. The second Lombard patent of 1907 described an improved paddlewheel engine, but these claims for the traction belt, the steering mechanism, sprocket wheels, and pivot shafts did not anticipate the features incorporated in the Holt patent of 1907. (12)

Pleased with this legal opinion, but aware of the uncertainties of patent law, Benjamin Holt pursued his investigations further. In the winter of 1909, he stopped in Washington, D.C., to consult the well-known authorities on patent rights, Pennie, Goldsborough & O'Neill. In particular, he asked them: first, did he have to own the Lombard patents in order to conduct his business secure from the claims of competitors; and second, if Holt purchased the Lombard patents, would this give him a substantial monopoly in the field of track-type traction engines.(13) The answers to Holt's questions were critical to his business future. If Lombard's patents were valid, Ben Holt at best would be liable for royalty payments, and at worst could lose his shirt. On the other hand, if the Holt patents were iron-clad, he might create a monopoly and freeze out all competition in the tread-tractor field.

In January 1909, Pennie, Goldsborough & O'Neill produced a lengthy report based on a careful analysis of the two Lombard patents. It concluded that the claims of the Lombard patents were of limited value because they could not be sustained in view of the prior art. Therefore, if Benjamin Holt bought these Lombard patents, they would be of little value in protecting his company in any court cases. As a result Holt made no further effort to secure the Lombard patents. (14)

Turning to the Holt patent of 1907 for the track-laying tractor, the patent experts concluded that virtually all of its 36 claims had been anticipated by a score of earlier patents and that "many of the supposed novel features of the Lombard and Holt traction engines were not novel with either of the inventors." As far as Holt's work was concerned, they reported, "None of the claims are of sufficient scope to insure . . . anything like a substantial

Alvin O. Lombard of Waterville, Maine was the first manufacturer to build engines with crawler that which were produced and sold in quantity. In 1900 he built his first steam engine designed very much like a railroad locomotive, except that it had sled runners designed for moving over snow in the lumber camps. His patent of May 21, 1901 featured a main frame to support the boiler and engine, and a supplementary frame to hold the sprocket wheels for driving the link-belt chain treads. The weight rested on rollers running inside the endless chain tracks. The frame holding the tracks was attatched to a large shaft which pivoted in a journal box. This provided flexibility for the machine in crossing rough terrain. The essential features of the 1901 patent were very similar to the Warren P. Miller engine of 1858.

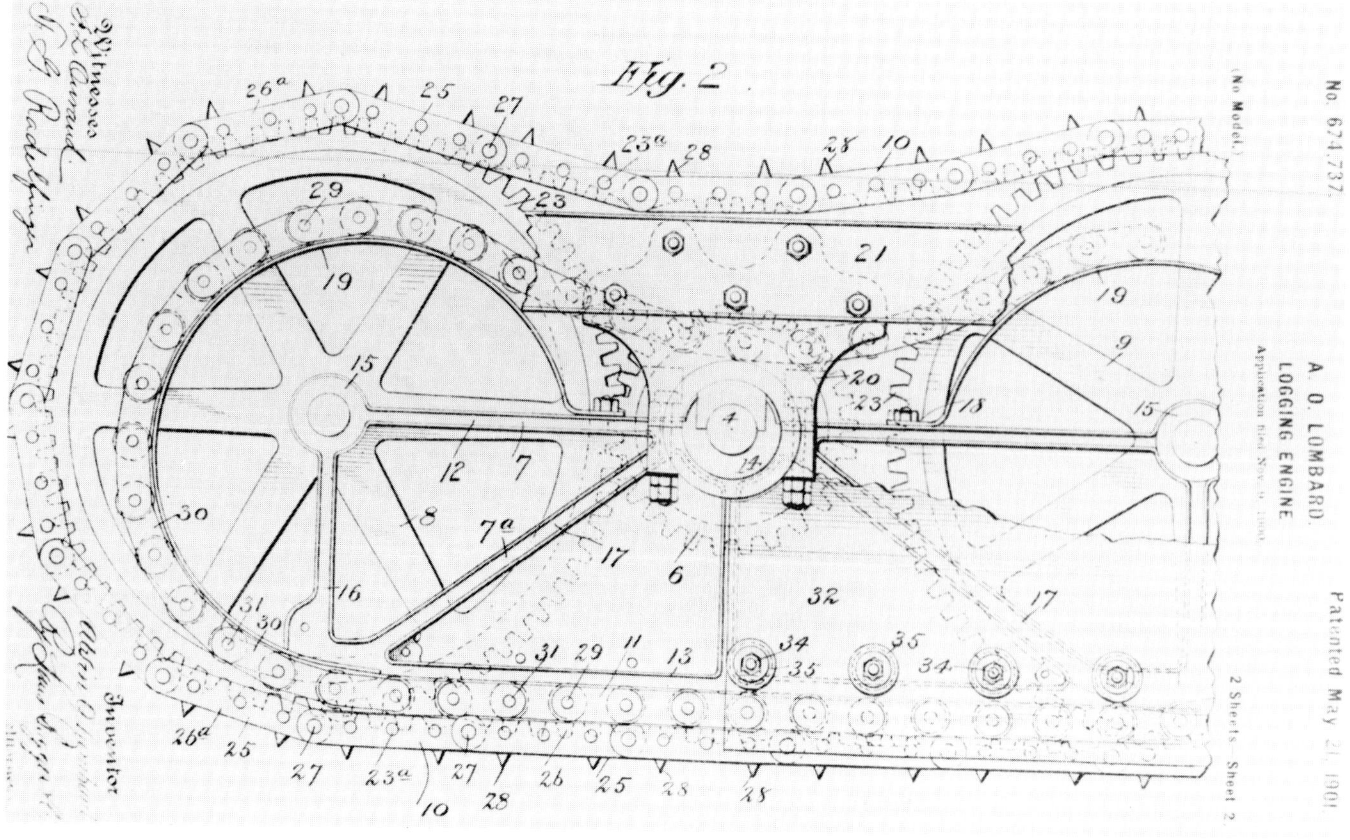

monopoly in the manufacture of traction engines of the caterpillar type."(15)

Having secured the best legal advice in the country, the Holts reacted calmly when Alvin O. Lombard visited Stockton in the winter of 1910 to complain about patent infringements by the Holt company. Such charges, they knew, would be difficult to prove in a court of law.

At the same time, the Holt company had spent up to $1 million developing its track-type tractors, and it understandably resented the attempt by a competitor like C. L. Best to appropriate its pioneering work without paying royalties. The company's only weapon was legal action. Even if the Holt patent documents were weak, no one could fully predict how the jury members would cast their ballots.

Accordingly, the Holts pressed their case against the C. L. Best Gas and Traction company with vigor. Their attorneys Charles E. Townsend and John E. Miller of San Francisco filed a suit in the United States District Court in February 1915 charging Best with violating 35 claims stated in the Benjamin Holt patent of December 17, 1907. When a defense motion to dismiss this action was denied in April 1915, both sides settled in for a long hard fight in the courts. (16)

C. L. Best, of course, had not been idle on this front, and he too had explored the legal ramifications of the existing patents. As early as 1907, he had hired attorney, E. L. Thurston, to give an opinion on the merits of the Lombard and Holt patents. Thurston concluded that both Holt and Best had "drawn upon the voluminous prior art."(17)

In preparation for the trial, both sides proceeded to gather a mass of evidence. Representatives of the Holt Company tracked down the whereabouts of each of the Lombard log haulers, and depositions were taken from Maine to Idaho.

Knowing that Holt's patent of 1907 pre-dated the first C. L. Best tractor patent of 1913, Best's attorneys J. J. Scrivner and Henry C. Montgomery of San Francisco reasoned that their strongest defense could be built around the contention that Holt had violated the Lombard patents, thus making any patent claims against C. L. Best void. To build a better case, Montgomery traveled to Waterville, Maine, in November 1915 to enlist Alvin O. Lombard as a friendly witness for the defense.

Lombard had carried a grudge against Ben Holt since 1910 when he had gone to Stockton to ask Holt to pay him royalties for patent infringements. Ben had showed Lombard around the factory and driven him in the country in his Oldsmobile roadster, but when the matter of money came up, Ben suggested merely that they divide the country; Lombard should stick to his log-hauling in the northern woods, and the Holts would take care of farming in the rest of the nation. According to Lombard, Ben had promised to write letters with more details, but he failed to do so. Therefore, when Henry Montgomery arrived in Maine in November 1915 offering to pay the railroad fare for Lombard and his daughter to attend the Panama-Pacific International Exposition and to be a friendly witness for the defense in the case, Lombard replied, "By God, young man, I'm glad to see you. If God Almighty could charter me to kill a man, I'd get on the train and go to California and kill old Ben Holt." (18)

Upon arriving in San Francisco, Lombard, then 60 years old, testified that he had built his first steam-driven log haulers in 1900, had improved them in 1905, and had granted a license to manufacture them to the Phoenix Manufacturing Company at Eau Claire, Wisconsin, for a royalty fee of $1,000 per engine. He estimated that he had manufactured almost 200 of these track-laying machines between 1902 and 1915, and that they were in use in every lumbering region from Maine to the Pacific Northwest.

When questioned about his conversations with Benjamin Holt in February of 1910, Lombard explained that he had asked Ben Holt for payments of royalties but Holt brushed aside the idea. Lombard said Holt admitted he had studied his logging machines in Montana before he began building his own track-type tractors. The Best defense also brought in C. H. Richardson, who had operated the Lombard log haulers beginning in 1905. Richardson testified that he had met Ben C. Holt in Missoula, Montana, in March 1906. Ben C., accompanied by his uncle Benjamin, he asked to see his Lombard engine, which they did, before staying the night with the Richardson family and continuing on their way. (20) This information was calculated to suggest that the Holts had pirated the Lombard system of building track-laying machines.

Throughout 1916, the Holt organization gathered information from the owners of Best tractors and, in addition, expanded their legal staff until it included a roster of nationally prominent attorneys. Never before had a more illustrious group of legal talent been secured for a lawsuit in the farm machinery industry.

Litigation in the case took a new twist when Best attorney Montgomery informed C. L. Best that since the Holt patents pre-dated those belonging to Best, the only way to gain an advantage would be to purchase the two Lombard log-hauling patents which pre-dated the first Holt patent for a track-type tractor. Best agreed and Montgomery took the train for a second time to Waterville, Maine, where Lombard agreed to sell his patents to the Best Company for $25,000. Lombard and Montgomery then traveled to Eau Claire, Wisconsin, where the two persuaded the Phoenix Manufacturing Company to cancel the Phoenix-Lombard royalty contract of 1904. Having exluded the Phoenix Company from responsibility for any future claims for damages—or shares in future spoils—Montgomery then filed a countersuit against the Holts for $3 million, claiming the Stockton firm had infringed the Lombard patents of 1901 and 1907 which now belonged to the C.

L. Best Gas Traction Company. (21)

By these astute moves, Montgomery threw the Holt forces on the defensive. With the scope of the trial broadened, additional testimony and affidavits would be required. More arguments would be made, and an appeal would be possible. The case might work its way to the Supreme Court.

Caught in this legal quagmire, the Holt people began assessing the future. If the case went to trial, at least 70 more witnesses would be called, some from as far away as Maine. Court sessions could last another six months. Already the case had required the constant attention of the company officials, with serious consequences for the ongoing company business. With the signing of the Armistice, it became imperative that the Holt efforts again concentrate on building the commercial farm machinery business which had been largely neglected during the war.

Faced with these unpleasant realities, the Holt and Best officials, prompted by their bankers, patent attorneys, and general counsel, began negotiations which eventually led to a compromise settlement signed on December 7, 1918. (22) The Holts received the satisfaction of a court ruling which agreed that their patents had been infringed on 12 specific counts, the basis of the 1915 suit. They also were allowed to purchase all the basic patents for the manufacture of track-type tractors, and accordingly they had the authority to demand from all other builders of crawler tractors royalty payments rangine from $25 to $100 per tractor, depending on engine horsepower. To both parties' benefit, the agreement banned any further suits between the Holt and Best companies, ending a 14-year feud which had probably cost the Holts alone about $750,000. (23)

The Best team also was gratified by the results of the settlement. They had viewed themselves as David against the Holt Goliath, a small company with $70,000 in capital taking on the giant Holt corporation with $10 million in assets. Best's lawyer Henry C. Montgomery later boasted that he and his 80-year-old partner J. J. Scrivner had successfully challenged the most powerful legal firms in the nation and fought them to a standstill. Whereas C. L. Best had told Montgomery as the final settlement began that Best would be satisfied to receive $50,000 and a license to continue to manufacture his tracklayers, his young attorney came back to San Leandro after four days of meetings with an award of $200,000 for sale of the Best patent rights. (24)

In the long run, both sides appreciated the truce because it reduced the constant financial drain of legal costs. The secretary of the Holt Manufacturing Company reported that the expenses incurred by the Holts in the patent department between 1915 and 1925 were $232,000, while the cost of litigation was $416,000. Not even a sizable company could afford to keep that kind of litigation going on forever. (25)

With the Best rivalry settled forever, the volume of Holt Company litigation decreased. Throughout the 1920s the main concern of Holt attorneys was the proliferation of other companies producing track-laying machines, many of which infringed on the Holt patents without paying royalties. As a result, claims were filed against such firms as the Trundaar Tractor Company of Anderson, Illinois; the Monarch Company of Watertown, Wisconsin; the Bates Tractor Company of Joliet, Illinois; the Bean Track-Pull Tractor Company of San Jose, California; and the Yuba Manufacturing Company of Marysville and Benicia, California. None of these claims resulted in major court cases. (26)

In retrospect it appears that the courts had only limited success in establishing the primacy of inventions associated with the development of the track-type machines built in the United States. With over 100 patents involved, they couldn't decide which one was the most original and constituted, in legal terms, the pioneer invention which in fact became the prior art. Even the best legal minds disagreed on the facts as well as the interpretations.

From an historical standpoint, however, the main honors go to Alvin Lombard and Benjamin Holt for first developing the new technology on a practical commercial basis. Had the Holt brothers not been involved, it is doubtful that the Lombard locomotives would ever have been suitable for farm tractor work. This remained Holt's outstanding contribution to world history—the development of the first successful track-type farm tractor, which later served as a prototype for the military tan, the bulldozer, and other track-laying machinery.

Luther Burbank, American naturalist at controls of a Holt two-ton tractor in 1924.

Chapter 18

CATERPILLAR AROUND THE WORLD

The settlement of the Holt company lawsuits in 1918, the death of patriarch Benjamin Holt in 1920, and the consolidation of the Holt and Best companies in 1925 proved to be major landmarks in the history of the famous Stockton company. The passing of the company's founder marked the end of the era of family control of the business and made way for the institution of modern corporate management practices. By pooling the resources of the two companies and eliminating the need for litigation aimed at maintaining a competitive edge in farm machinery production, the merger laid the basis for sound economic growth in the late 1920s.

From 1925 to 1930, Caterpillar Tractor Co. sales jumped 229 percent from $13.8 million to $45.4 million, and profits climbed from $3.3 million to $9.1 million. The number of company shareholders increased from 1,919 to 12,812, and employment rose from 2,537 to 6,282 workers. (1)

Opening the door to the new era was the good news in 1926 that the War Department had designated Caterpillar "60", "30", and "2-ton" tractors as the official tractors of the United States Army. Caterpillar Tractor Co. President R.C. Force thanked Army officials for this special recognition and asked if his company could publicize this information in new company advertising campaigns, a request which was granted.(2)

Another new publications campaign calculated to display the growing diversity of the new Caterpillar products resulted in the popular publication, "The Romance of Caterpillar Tractors". This brochure announced that bore and stroke today took the place of blood and brawn. Men and mules were emancipated—the steel sinews of the tractor uncomplainingly took up the burden of tired muscles.

Caterpillar crawlers, the bulletin suggested, could work independently of the weather and make enough money to afford a high school education and music lessons for farmers' children, a radio for their homes, and an occasional trip to town to see Hollywood movies.

Larger than life were these machines that performed heroic deeds in Caterpillar promotional literature of the new era. A man suffering a heart attack in Yellowstone National Park was rushed through deep snow to a hospital by a giant snow-moving Caterpillar track-type tractor, just as a second crawler hauled feed to blizzard-isolated sheep in Wyoming. Another heroic tale had two Caterpillar "60s" fighting a raging wheat field fire and saving a $100,000 crop.

Integral to the Caterpillar Company's expansion in the 1920s was the search for new applications of the technology pioneered by Benjamin Holt in the preceding four decades. Road-building machinery became a major focus as the nation's highway system began to grow with the infusion of state highway funds. In 1928, Caterpillar acquired the well-known Russell Grader Manufacturing Company of Minneapolis, and three years later Caterpillar introduced its first self-propelled road grader. Caterpillar crawler tractors and road machinery were used by the Cuban government in this same era to build that country's Central Highway spanning the island. The Cubans were so proud of this project that they sent an exhibit featuring a model of a Caterpillar "60" to Seville, Spain.

The leading figure in the development of modern earth-moving machinery, however, was another Valley man, Robert G. LeTourneau, who got his start leveling land near Stockton in 1919. Recalling the satisfying feeling of being able to move land, LeTourneau remembered how he began thinking about making improvements on existing Holt-made earth-movers:

> When I was in dry going, the track-type treads of the Holt threw sand in my face. When I was down in the bottoms, they threw mud, and I loved it. When I saw a hummock ahead to be cut down, I charged it as though I was a knight in a tournament, and if my scraper bounced off without slicing off a good cut, I was furious. I wanted to move dirt. Lots of dirt.

After considerable experimentation, LeTourneau attatched two 10-horsepower electric motors on a scraper, one at each end of the blade, and ran wires to a switch near the tractor seat. The result was a one-man outfit where the driver of the tractor could operate the scraper by simply pushing a few buttons with his fingertips.

During the 1920s LeTourneau manufactured his earth-moving scapers in Stockton where they were sold for building roads and laying water mains. In the early 1930s LeTourneau teamed up with industrialist Henry J. Kaiser on contracts for the then-largest building project in the world, Hoover Dam on the Colorado River.

In 1935, LeTourneau followed the example of the Holt Manufacturing Company by moving his factory off the far West Coast to the centrally located city of Peoria. Here he built huge scrapers capable of carrying 50 tons of dirt at a speed of 15 miles per hour.(4) In years since, his widely recognized giant machines have been used in constructing dams, moving missile carriers, and erecting off-shore drilling platforms around the world.

The expansion of America's heavy machinery industry including the work by Caterpillar and LeTourneau, was cut short by the stock market crash of October 1929 and the advent of the Great Depression. Although economists are still debating the complex causes of this debacle, in part they believe it stemmed from the failure of the prevailing laissez-faire economic gospel which held that the large profits by unregulated big businesses would "trickle down" to the workers below and keep the economy healthy and growing. However, while the productivity rate of workers increased 33 percent during the 1920s—in large part because of the mechanization of industry—factory wages remained constant. As for the American farmer, his income averaged a lowly $548 per year, despite a quadrupling of the nation's wealth between 1900 and 1929. When almost three-fourths of the nation's families found themselves living on incomes of under $2,500 a year, their severely decreased purchasing power resulted in a stockpile surpluses in warehouses. Then the nation's mighty industrial machine ground to a halt.(5)

By 1932, farmers' incomes reached the lowest ebb since the time of George Washington, and wheat sold for as little as it did during William Shakespeare's era. In the American Midwest, farmers found themselves unable to pay their debts with oats selling at six cents a bushel, corn at 12 cents a bushel and hogs at two cents a pound. While farmers were being ousted from their homesteads, industrial kings like Henry Ford, for all their business expertise, could only suggest that President Herbert Hoover clear the way for what Almighty God was going to do.(6)

Dependent in large part on sales to farmers, the Caterpillar Tractor Co. saw its profits of over $12

"Caterpillar" Diesel Fifty Tractor with John Deere 3-12' weeders, weeding newly seeded wheat land seeded with deep furrow drills. Weeder is knocking out weeds on top of ridges between furrows and not hurting the young wheat. Caterpillar Tractor Co. photograph.

Armed bodyguard with Caterpillar Diesel D4 Tractor with armored bulldozer of Engineer Combat Regiment on Munda Air Field. New Georgia Island in the Solomons. Caterpillar Tractor Co. photograph.

million dollars in 1929 turn to a loss of $1.6 million in 1932. In response, the company retrenched by laying off half of its 6,875 factory workers.(7)

Some Caterpillar dealers, such as Benjamin's son William K., began selling road-building equipment to county governments, that could only offer warrants rather than cash. His partner, T.R. Finley, sold machinery for gold and silver in Mexico, transferring this specie to Canadian banks because trading in gold had been banned in the United States.(8)

The sale of Caterpillar crawlers to farmers, of course, slowed to a near halt as the Depression wore on. In Texas, cotton went unpicked because workers could earn only 70 cents a day, while 70 percent of the farmers in Oklahoma were unable to pay interest on their mortgages. Roads and railroads in the Southwest teemed with ragged hitchhikers.(9)

Even nature turned against farmers in the Midwest. In this region where 20 inches of rainfall annually was required to raise grain, a prolonged drought set in. By 1933, top soil began blowing into the air, eventually causing 50 million acres to be blighted by dust storms extending from Canada to Mexico. Animals choked to death from the dust, and on April 10, 1935, members of the Texas legislature wore gas masks in the state house.(10)

With only a few people able to purchase new machinery during these catastrophic years, over half the tractor companies in the nation went bankrupt. Only foresight allowed the Caterpillar Tractor Co. to survive the ravages that brought other companies down. In 1931, Caterpillar Tractor Co. introduced the nation's first successful diesel-engine tractor, an attractive new product that became a timely source of income for the corporation. By 1933, at the very depth of the Depression, new orders come in for diesel tractors and when the nation's total diesel production hit 2,000,000 horsepower in 1937, Caterpillar machines accounted for one-third of this energy bank. Accordingly, total Caterpillar sales climbed from $13 million in 1932 to $63 million in 1937.(11)

The diesel engine began with the work of Rudolph Diesel, a German engineering student. Hoping to improve the low efficiency of the steam engine, he built his first "diesel" model in 1892, which almost killed him when he tried to start it. During the next several years he improved the design, and the first diesel engine used commercially in the United States was operated in St. Louis in 1898.(12)

Within a few years, thousands of heavy stationary diesel engines were in use under conditions where they operated at uniform speed under uniform loads. Their great advantage over the traditional gasoline engine was their ability to extract more energy from the same amount of fuel. As yet, diesels were not suited to run tractors because the engines were too heavy and lacked flexibility in producing a wide range of power. Producing nearly twice as much work as a gasoline engine from a comparable amount of fuel, diesel engines were also much cheaper to operate because they ran on a special low-grade fuel, a distillate similar to that used in home oil furnaces.

At the heart of the diesel engine is the fuel injection system. Each engine cylinder has a pump which forces the fuel oil through a nozzle fitted with fine holes. As the fuel is forced through these holes at great speed, it breaks up into a fine spray which ignites immediately when subjected to intense heat and pressure in the cylinder head. This ignition occurs without the use of a spark from a spark plug because the high compression ratio of at least 16 to 1 ignites the fuel vapor instantly.

Although diesel motors were widely used from 1898 to 1931 in the United States, they still had to be refined for mobile use, a process which required nearly 30 years of experimentation. The most difficult problems associated with the engine were the building of pumps accurate enough to force fuel into a cylinder containing 600 pounds of compression and the construction of a nozzle which would produce a spray of fuel fine enough for instant spontaneous ignition.

Pliny E. Holt spent many years trying to improve the diesel engine, and the Caterpillar Tractor Co. invested $1 million in diesel research from 1926 to 1931. The company secured information from Europe and brought a large collection of Diesel motors to San Francisco for study. Caterpillar engineers worked for several years on the difficult task of designing a pump with the fine tolerances needed to enable the machine to work. So snuggly did the final pump cylinder fit the piston that when a mechanic held it in his hand, his body warmth would expand the piston enough to prevent it from entering the pump cylinder. The temperatures of both the piston and cylinder, then naturally had to be equal before the parts would function.(13)

By 1931, Caterpillar engineers had finally designed diesel engines which operated smoothly and efficiently under a wide variety of work loads. Their engines were free from delicate adjustments, had working parts encased in dust-proof casings, and were constructed to be easy to service and operate.

Just as the Holt Manufacturing Company had introduced the combined harvester, the steam traction engine and the gasoline tractor to the buying market with public demonstrations of their merit, so the Caterpillar company planned numerous public showings of new diesel tractors to convince a skeptical and financially strapped buying market that the machines were practical and worth investing in. One demonstration in Oregon in the spring of 1932 required a Caterpillar Diesel crawler to run 23 hours daily in order to plow 6,880 acres in 46 days, at a cost of only eight cents an acre. The "Oregon Farmer" commented, "As the group of 150 men from half a dozen states watched...in the closing hours of one of the most spectacular performances in the history of

A photograph from around 1910 shows the main portion of former Colean plant in East Peoria, Illinois. Holt Caterpillar Company ceased to be a separate entity 1913, when it was incorporated into the Holt Manufacturing Company. The favorable price of this Colean factory building and equipment proved to be one of the most important factors in establising the Holt Caterpillar Company in Peoria.

big-scale farming, some of them had the thought in mind that they were present at an historic event...a distinctive milepost in the long trail of agricultural history." (14)

Additional publicity for the Caterpillar diesel machines came when William Hazlett Upson began writing short stories about the exploits of the drivers of track-type tractors. The stories' principal character was Alexander Botts, an energetic and imaginative salesman for the Earthworm Tractor Company, who engaged in battle of wits with sales representatives from the rival "Behemoth" wheel tractor company. This fierce competition led to diabolical schemes by enemy salesmen to discredit Botts and his Earthworm tractors, resulting in dangerous situations over which Botts eventually triumphed.

Upson's stories appeared in the "Saturday Evening Post", with illustrations by E. F. Ward, over a period of ten years prior to the outbreak of World War II. In "The Case of the Lost Overalls", which appeared in September 1941 "Post", Botts received orders from the Los Angeles Earthworm Tractor Co. to sell tractors to "Vesuvius" Johnson, a fiery contractor building a road through a nearby canyon. Botts drove one of his tractors to the construction site and had to sell it to Johnson, "a veritable wild man of the mountains." Johnson decided Botts was a nuisance and ordered him off the road. After Botts refused, Johnson ran the Earthworm tractor over the cliff, where it landed in the still soft concrete pillar of the bridge under construction. The impact broke off the pier, and both pillar and tractor tumbled into the canyon floor several hundred feet below. When a crane later lifted the Earthworm from the canyon floor, it was discovered to be undamaged, causing "Vesuvius" Johnson to explode and claim that Botts had somehow switched tractors. Botts proved the machine to be the original one by pulling Johnson's overalls out of the machine, and a chastened Vesuvius bought 21 Earthworm tractors to replace his inferior Behemoth machines.(15)

In 1936, the perils of Botts caught the attention of motion picture producers in Hollywood who made a movie about him called "Earthworm Tractors," starring comedian Joe E. Brown. Warner Brothers Studio cameramen filmed location shots in the Peoria Caterpillar factory.(16) In 1937 Warner Brothers released a second movie, "God's Country and the Woman," set in the beautiful timber region of Southern Washington. Caterpillar diesel crawlers were featured performing difficult lumbering operations, while actor George Brent busily dynamited log jams and fought free-for-all battles with loggers from rival lumber camps.(17)

During the Second World War, the Caterpillar company became heavily involved in producing war materials. A few months before the attack on Pearl Harbor in late 1941, the Army Ordnance Department requested the Peoria officials to manufacture for the M-4 tank an air-cooled diesel motor which could burn heavy oils and low octane gasolines as well as the conventional distillate. These engines began rolling off the assembly line in July 1943. In addition, the Peoria factory produced 155-millimeter howitzer carriages, shells, bomb parts, and track-type mechanisms for military vehicles.(19)

Because of the belief that the Japanese might invade the North American continent via Alaska, U.S. Government officials decided in early 1942 to construct the Alcan Highway. This 600-mile road stretching from Fort St. John in British Columbia to Fairbanks, Alaska, would allow the United States to ship material and supplies by land to Alaska. Brigadier General William M. Hoge was placed in command of this project and established headquarters in the tiny town of Whitehorse. Men, machines, and supplies were mobilized and rushed to the scene in a frenzied effort to construct the highway without delay, a job completed by 10,000 soldiers and 6,000 civilians in slightly over six months. Pushing forward at a rate of 8 miles a day, the crews bridged 200 streams, laying a roadway 24 feet wide across mountain ranges with elevations of 4,000 feet. Nearly 70 percent of all the tractors used on this remarkable project were Caterpillar-built. (20)

A vivid account of this epic feat in "The National Geographic Magazine" in February 1943 described the hardships of building the road and contending with ice and snow, clouds of mosquitoes, and the formidable terrain. Author Froelich Rainey described 20-ton track-type tractors equipped with cutting blades advancing into standing timber to push aside trees, stumps, and rocks as if the forest growth was no more than cornstalks. Like "wild boars rooting and snorting in the jungle," the machines were driven by operators who had only a heavy steel bar running over the length of the tractor for protection against trees that sometimes fell on the machines. Some tractors worked half-submerged in swamps and muskeg; at midnight the throb of tractors and the roar of trucks could be heard. As one crew replaced another around the clock, disabled tractors were serviced at day and night motor pools. As a result, the highway was completed in November 1942 in record time.

During the war years, 85 percent of the productive capacity of the Caterpillar Tractor Co. went to agencies of the Federal Government. To increase the manufacture of war materials, employment at the company's plants grew from 11,000 in 1940 to 20,000 in 1944. When 6,000 of these workers went off to war, women entered the factories to comprise 30 percent of the payroll.(21)

As the war raged on, the "Cat" bulldozer became known as the work horse of the Allied effort. This machine featured a blade mounted on the front of the tractor which could be raised or lowered by cable systems. In the Pacific campaign, the island-hopping tactics employed by the American forces required large numbers of bulldozers for clearing airfield landing strips, removing coral reefs, throwing up earthworks, and building roads. The machines were so important that they were frequently given landing priorities before tanks and other weapons of war. Calling them the "Boss of the Beach," Major General Eugene Reybold described them as an indispensable all-purpose weapon used to dig trenches, destroy tank traps, fill bomb craters, position pontoon bridges, tow aircraft and power generators for producing electricity in advance bases.(22)

The history of the Caterpillar Tractor Co. in the post-war years has been one of consistent growth. Under the leadership of Louis B. Neumiller and Harmon Eberhard, both former Holt Co. employees, the company expanded production. Several new factories were built in Peoria and other plants were constructed in York, Pennsylvania, and in Joliet and Decatur, Illinois. By 1955, over $200 million were spent in this expansion project.(23)

The assembly lines in the 1950s produced more powerful tractors such as the D8 model which weighed 38,155 pounds and delivered 150 horsepower at the drawbar. The DW 21 tractor could scoop up 20 cubic yards of dirt and carry it off at 20 miles an hour. Auxiliary equipment powered by Caterpillar crawlers included pipe layers for use in the petroleum industry; tractor shovels for loading logs, coal and dirt; and motorized road graders and scraper units for highway construction. By 1963, the company was producing 141 different products.

In more recent years, Caterpillar products have become a common sight in most countries of the world. As the result of the export of modern machinery and scientific information, many nations' food supplies have increased dramatically. In Pakistan, for example, the United States assisted in building the Mangla Dam, a billion dollar project to utilize the water of the Indus River for irrigation. Caterpillar supplied seven million dollars worth of crawler tractors for moving dirt at the site, where the completed dam provides water for farms that now feed millions of people.(24)

Similarily, the goverment of Australia, Thailand, Japan, Malaysia, Indonesia, and Mexico have purchased track-type tractors, motor graders, bulldozers, and diesel engines to help people achieve a higher standard of living. The machines have built irrigation canals in India and roads in Borneo, cleared land in Malaysia, logged in the Philippines, extended rice paddies in Japan, and moved dirt in Mexico.(25)

In Brazil, bulldozers are working on the Trans-Amazon Highway, which will stretch 3,100 miles from eastern Brazil to the Peruvian border. This road which will help eliminate the isolation of rural Brazilians, will give them access to city markets.

During the last few years, a United States development program near Seoul Korea, has used Caterpillar machines to return abused farmland to rice paddies, irrigated fields and fish ponds. Terraced hillsides reduce erosion, and farmers are learning the advantages of crop rotation, animal husbandry and community cooperation.

Today, something of the magnitude of the productive capacity of the Caterpillar Tractor Co. can be seen in the corporation's annual reports. Total sales rose from $4,963.7 million in 1975 to $9,154.5 million in 1981—an increase of almost 100 percent. Foreign sales in 1981 represented 56.6 percent of total sales, thus giving Caterpillar's business activities a $3.3 billion net favorable balance of trade for the United States.

The story of the evolution of technology that lightens the burden of labor among the people of the world and raises their expectations for a better standard of living is a saga of talent, dedication, and optimism. It is difficult to find people more instrumental to this significant advancement than the Holt families, especially California's preeinent agricultural industrialist, Benjamin Holt.

A Holt "75", gasoline track-type tractor built about 1920 at Stockton is one of the centerpieces of the Holt Memorial Hall in the Haggin Museum. Weight, 23,600 pounds; width, 8 feet; length 20 feet; height 10 feet tracks 6 feet 3 inches; brake horsepower 75; drawbar horsepower 50 Motor 4-cylinder, valve-in-head; bore 7½, stroke 8 inches, speed 550 r.p.m. Fuel tank, 74 gallons; water tank; 53 gallons

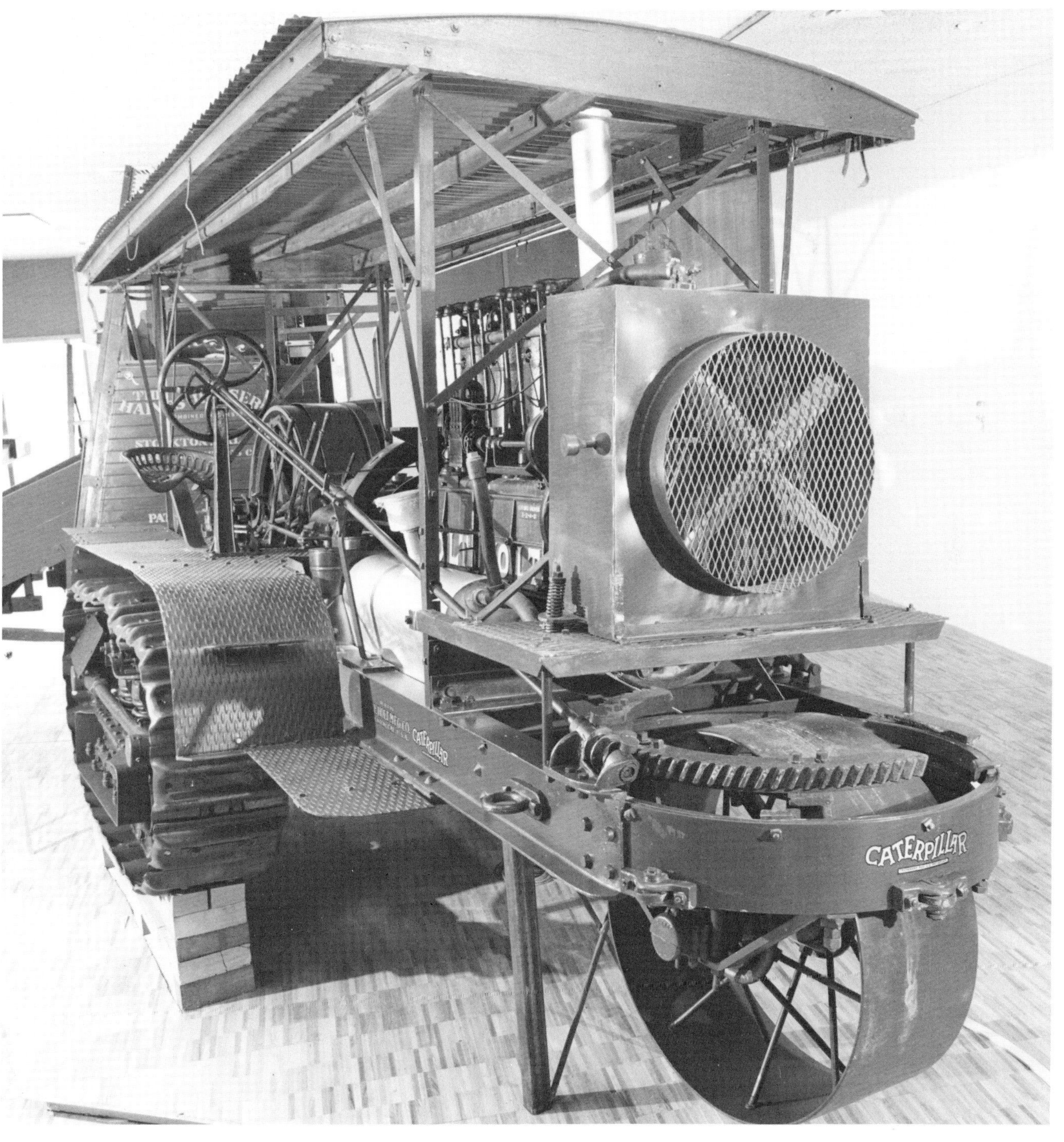

Significant developments and locations are recognized by the American Society of Agricultural Engineers as Historic Agricultural Engineering Landmarks to serve as benchmarks in the progress of engineering applied to agriculture and to encourage today's inventors to develop new solutions to problems in food and fiber production and processing systems. The two ASAE Historic Landmarks dedicated on January 13, 1983 at the Haggin Museum's Holt Exhibit are the sixteenth and seventeenth landmarks designated by ASAE.

ASAE President Bill Harriott and Holt Brothers descendant Harry Holt at the ASAE Historic Landmark dedication on January 13, 1983.

The first ASAE Historic Landmark at the Holt Exhibit was awarded for "The Sidehill Combine", developed and patented by the Holt Brothers at Stockton, California in 1891.

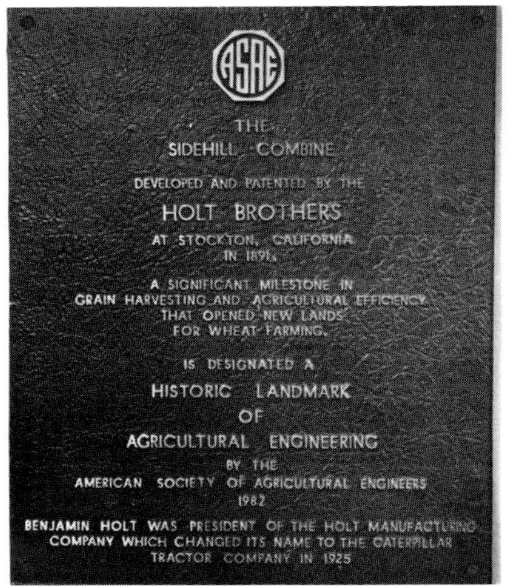

The second ASAE Historic Landmark at the Holt Exhibit was awarded for the first successful "Track-type Tractor", developed by Benjamin Holt at Stockton, California in 1904.

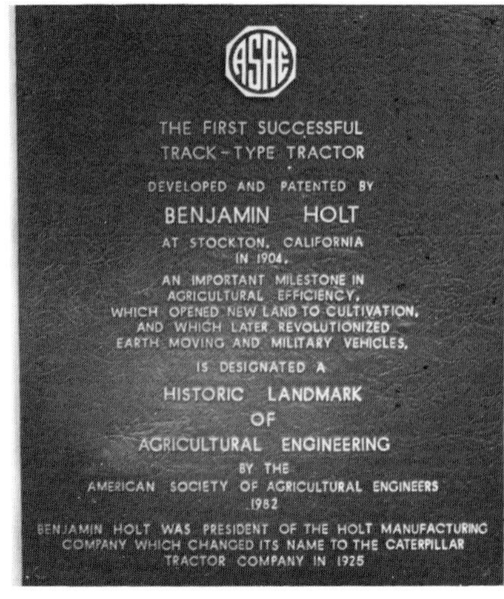

FOOTNOTES

CHAPTER 1

1 The most comprehensive primary source materials pertaining to the life of Benjamin Holt and the history of the Holt Manufacturing Company are located in the Library of the Pioneer Museum and Haggin Galleries in Stockton, California. This collection includes business records, correspondence, legal documents, scrapbooks, newspaper clippings and hundreds of photographs.

The recent acquisition of the "Pliny E. Holt Papers" consisting of two filing cases of the business correspondence of the executives in the Holt Company during the years 1900 to 1930 constitutes an extremely valuable body of material because the information illuminates the events surrounding the development of the first Caterpillar track-type tractors, and the decisions to establish a branch factory in Peoria, Illinois in 1910. These papers have been in the custody of Frank Harrison, and Pliny G. Holt in Potomac, Maryland.

The second best source of historical materials relating to Benjamin Holt and all of his works is the F. Hal Higgins Collection of Agricultural Technology located in the Library at the University of California at Davis. F. Hal Higgins spent half a century collecting information about farm machinery and his collection now numbers 300,000 items including agricultural journals, manufacturers catalogues, calendars and posters, and thousands of photographs.

2 F. Hal Higgins, "The Story of Track-Type Tractors from 1713 to 1904." 14-page manuscript. F. Hal Higgins Collection. Shields Library, University of California, Davis, California.

3 "Fifty Years on Tracks,"(Peoria, Illinois: Caterpillar Tractor Co., 1954), page 53.

4 "Great Register of San Joaquin County,"(Stockton, California: City Public Library). This record includes the list of voters and their age and personal description.

5 Interview with William K. Holt, Hillsborough, California by Raymond Hillman and Reynold M. Wik, June 6, 1975.

6 Interview with James K. Page in Alemeda, California by Reynold M. Wik, November 24, 1975. Mr. Page's father was Benjamin Holt's personal assistant which included driving his automobile.

7 Caterpillar Tractor Co. Business Records, Legal Files of the Patent Division. "Patents and Trade Marks," vol. I., Index. Main Office Building, Caterpillar Tractor Co., Peoria, Illinois. The patent files listed include Benjamin Holt, 47 patents; Pliny E. Holt 38 patents; E. E. Wickersham, 42 patents; Ben C. Holt, 2 patents; William Turnbull, 35 patents; E. F. Norelius, 20 patents and Henry T. Preble, 3 patents.

8 Interview with Warren H. Atherton of Stockton by Reynold M. Wik, June 20, 1975.

9 Carl Crow, "Uncle Ben and the Caterpillar,"*Sunset*, November 1917, pages 21-22.

10 Reginald T. Townsend, "Tanks and the Hose of Death,"*Worlds Work*, December 1916, page 196.

11 "The Granite Monthly," (Concord, New Hampshire), April 1921, page 2.

12 "Pacific Rural Press," (San Francisco, California), December 11, 1920, page 755.

13 Interview with Paul E. Weston in Stockton, by F. Hal Higgins, July 10, 1952.

CHAPTER 2

1 "The Three Generations of Holts in America,"(Newburgh, New York: The Holt Association of America), pages 1-3.

2 Personal letter written by Wallace K. Ewing, Dean of Colby-Sawyer College, New London, New Hampshire to Reynold M. Wik, Oakland, California, September 15, 1975.

3 Henry F. Pringle and Kathern Pringle, "Concord, New Hampshire," *Saturday Evening Post*, October 22, 1949, pages 32-35.

4 Interview with John Gravelle, Concord, New Hampshire by Reynold M. Wik, June 26, 1975.

5 J. M. Guinn, "History of the State of California and Biographical Record of San Joaquin County,"(Los Angeles, California, 1909), vol. 2, pages 75-76.

6 "Holt Brothers—Importers of Carriage Stock and Hardware,"(New York: Hub Publishing Company, 1880), page 14.

7 Interview with Pliny E. Holt, Stockton, California by F. Hal Higgins, June 20, 1930. F. Hal Higgins Collection.

8 "Concord Daily Monitor," (Concord, New Hampshire), March 18, 1873, page 4.

9 Interview with William K. Holt, Hillsborough, California by Reynold M. Wik, November 12, 1975.

CHAPTER 3

1 Hubert H. Bancroft, "The Works of Hubert H. Bancroft," (Santa Barbara, California, 1970), vol. 6, page 457.

2 "The Stockton Journal," Stockton, California, November 30, 1852, page 1.

3 Horace Davis, "Wheat in California,"*Overland Monthly,* November 1868, page 65.

4 Gilbert C. Fite, "The Farmer's Frontier: 1850-1880," (New York, 1966), page 156.

5 Hubert H. Bancroft, "History of the Life of Austin Sperry," (San Francisco, 1892), pages 548-554.

6 "Pacific Rural Press," September 7, 1872, page 153.

7 *Ibid.,* October 20, 1886, page 373.

8 Leo Rogin, "The Introduction of Farm Machinery in its Relation to Productivity of Labor in the Agriculture of the United States During the Nineteenth Century," (Berkeley, California: University of California Publications in Economics, 1931), vol. 8, page 40.

9 Robert Ferguson, "Benjamin Holt and the Holt Manufacturing Company," 5-page typescript, 1940, page 2. The Haggin Museum.

10 "Stockton Daily Independent," (Stockton, California), March 17, 1883.

11 Rodman Paul, "The Wheat Trade Between California and the United Kingdom,"*The Mississippi Valley Historical Review,* December 1958, page 397.

12 "Sales Reports for the Year 1910" Business Records, Holt Manufacturing Company. The Haggin Museum.

CHAPTER 4

1 "Caterpillar Times," (Stockton: Holt Manufacturing Company), May 1915, page 8.

2 William W. Dingee, "Early American Threshing Machines," *Thresher World and Farmer's Magazine,* May 1903, page 15.

3 Robert L. Ardrey, "American Agricultural Implements," (Chicago, Illinois, 1894), page 104.

4 Editorial, "Cultivator and Country Gentleman," October 18, 1877, page 664.

5 George Miller, "The Harvest Season,"*Indiana Farmer,* September 7, 1878, page 1.

6 "Pacific Rural Press," May 6, 1871, page 1. Also, October 26, 1946, page 331.

7 "Hebron Nebraska Journal," (Hebron, Nebraska), September 12, 1876, page 3.

8 F. Hal Higgins, "The Harvest Cook Wagon," 2-page typescript. F. Hal Higgins Collection.

9 Henry Domonoske, "Recollections of the Big Change from Stationary Threshers," 9-page typescript. F. Hal Higgins Collection.

10 F. Hal Higgins, "Our Centennials and Agriculture,"*California* June 1947, page 17.

11 F. Hal Higgins, "The Moore-Hascall Harvester Centennial Approaches,"*Michigan History Magazine,* 1930, vol. 14, page 419.

12 Lillian Church, "Partial History of the Development of Grain-Harvesting Equipment," U.S. Department of Agriculture, Information Series, No. 72. 1939, page 45.

13 "Oakland Tribune," (Oakland, California), January 2, 1955, page 10.

14 "California Farmer," September 19, 1862, page 24.

15 No. 408, 413, United States Patent Office, August 6, 1889.

16 "Pacific Rural Press," July 1885, page 468. Also, June 25, 1887, page 578.

17 *Ibid.,* page 468.

18 "Stockton Daily Evening Record," June 14, 1933, page 1.

19 George Ingersoll, "My Experiences With Early Combined Harvesters," 2-page typescript, 1927. F. Hal Higgins Collection.

20 F. Hal Higgins, "Tom Luke Recalls the California Combines," 22-page typescript, 1931. F. Hal Higgins Collection.

21 "Catalogue," (Stockton, California: Holt Manufacturing Company), 1917, page 2. See also, F. Hal Higgins, "97 Years of Combining in California,"*Calfornia Farmer,* March 25, 1950, pages 280-281.

22 D. W. Meinig, "The Great Columbia Plain," (Seattle, 1968), pages 392-404.

23 Miscellaneous notes. Business Records of the Holt Manufacturing Company. The Haggin Museum.

24 Mrs. Fred Franke, "Threshing Rings,"*Dakota Farmer,* October 1948, page 19.

25 Business correspondence, Holt Manufacturing Company. Letter, Martha Walgren, Modesto, California to Benjamin Holt, July 20, 1908. The Haggin Museum.

26 Interview with Melvin Reister, Williams, California by Raymond Hillman, November 12, 1974.

27 "Pacific Rural Press," October 4, 1890, page 298. Also, May 13, 1916, pages 585-589.

28 "Adventures in Combining," 2-page typescript, 1937. F. Hal Higgins Collection. Also, Interview with Paul E. Weston by F. Hal Higgins and published in "I Was the Man Who,"*Iron-Men Album Magazine,* (Lancaster, Pennsylvania), September-October, 1953, page 11.

29 It should be noted that after 1900, the side-hill harvester mechanism was changed so that the combine drive wheels could be raised or lowered by means of a cylinder encasing a screw-shaft.

30 Tom Luke, "History of the Combined Harvester," 36-page typescript, page 28. The Haggin Museum.

31 Roy Bainer, "The Engineering of Abundance, An Oral History Memoir," 1975, page 449. Oral History Center, University of California, Berkeley.

CHAPTER 5

1 C. Parker Holt, "Early Developments of the Holt Manufacturing Company," 10-page typescript. The Haggin Museum.

2 "Stockton Evening Mail," December 24, 1904, page 10.

3 "Oakland Tribune," January 15, 1919, page 8.

4 "Pacific Rural Press," July 1891, page 465.

5 Margaret E. Robinson, "From Lumber and Grain to Tracks," 49-page typescript. May 15, 1963. F. Hal Higgins Collection.

6 Samuel Smiles, "Lives of Engineers, Boulton and Watt,"(London, 1904), page 303.

7 "House Executive Documents," 25th Congress, 3rd Session II, no. 345, pages 1-472.

8 "Farm Implement News," (Chicago, Illinois), December 1895, page 4. The first portable farm steam engines were built by A. L. Archambalt in Philadelphia in 1849.

9 Mary Dodge Woodward, "The Checkered Years," (Caldwell, Idaho, 1937), page 93.

10 "Pacific Rural Press," June 14, 1884. page 48.

11 Reynold M. Wik, "Steam Power on the American Farm," (Philadelphia, 1953), pages 251-257.

12 Lynn W. Ellis and Edward A. Rumely, "Power and the Plow," (New York, 1912), page 283.
13 F. Hal Higgins, "Condensed Chronicle of the Combine," *Implement Record*, July 1949, pages 22-23.
14 "Pacific Rural Press," August 2, 1887, page 114.
15 "Farm Implement News," (Chicago, Illinois), January 7, 1893, page 2.
16 "Pacific Rural Press," August 16, 1890, page 139. Also, August 24, 1889, page 149.
17 "Pacific Rural Press," August 24, 1889, page 149.
18 "Pacific Rural Press," August 2, 1887, page 114.
19 "Pacific Rural Press," August 16, 1890, page 159.
20 George L. Dickenson, "Diary, 1890-1909," The Haggin Museum.
21 "California Farmer," February 10, 1951, page 118.
22 "Logging and Farm Engineers,"*Engineers and Engines*, (Joliet, Illinois), May-June, 1969, page 2.
23 "Stockton Evening Mail," May 10, 1903, page 2.
24 "Pacific Rural Press," November 22, 1890, page 443.
25 Robert Ferguson, "Benjamin Holt and the Holt Manufacturing Company," 5-page typescript. May 2, 1942, page 5. The Haggin Museum.
26 Russell C. Grimsley, "Tales of California," (Columbia, California, 1966), page 117.
27 *Op. Cit.*, "Dickenson Diary," January 1, 1896; January 18, 1895; January 19, 1895; January 20, 1895.
28 F. Hal Higgins, "Ben Holt Gave Me Steam Tractor Lessons," *California Farmer*, August 26, 1950, page 161.
29 "Pacific Rural Press," July 16, 1887, page 45.
30 "Pacific Rural Press," July 27, 1889, page 71.
31 "Pacific Rural Press," July 19, 1890, page 58.
32 Interview with Clifford M. Griffen, Williams, California by Raymond Hillman, November 18, 1974.
33 "Pacific Rural Press," September 17, 1887, page 223.
34 "Threshermen's Review," (St. Joseph, Michigan), July 1898, page 28.
35 John Thompson, "The Settlement Geography of the Sacramento-San Joaquin Delta of California," (Doctoral dissertation, Stanford University, Stanford, California, December 1957), page 1.
36 "The Byron California Times," (Byron, California), 1910, page 2.
37 "Fifty Years on Tracks," (Peoria, Illinios: The Caterpillar Tractor Company), page 10.
38 *Op. Cit.*, Reynold M. Wik, "Steam Power on the American Farm, " pages 100-105.
39 W. C. Smith, "The Old Thresherman Talks of Quitting,"*American Thresherman and Farm Power*, (Madison, Wisconsin), October 1920, page 8.
40 "Thresherman's Review," May 1898, page 23.

CHAPTER 6

1 "An Agreement of Dissolution and Assignment," 2-page transcript, October 23, 1888. This document is in the possession of Frank Harrison Holt, (a son of Pliny E. Holt) at his home in Potomac, Maryland.
2 C. Parker Holt, "Early Development of the Holt Manufacturing Company," 22-page typescript, 1935. The Haggin Museum.
3 Philip S. Rose, "The Holt Manufacturing Company," 6-page typescript, January 20, 1915. F. Hal Higgins Collection.
4 Interview with William K. Holt, Hillsborough, California, by Reynold M. Wik, November 13, 1975.
5 *Op. Cit.*, C. Parker Holt, "Early Development of the Holt Manufacturing Company."
6 "Numerical Index of Patents," (Library, Caterpillar Tractor Co., Peoria, Illinois), December 31, 1931.
7 "Analysis of Sales. Block 1-16," 18-page report in the Pliny E. Holt Papers. The Haggin Museum.

CHAPTER 7

1 F. Hal Higgins, "The Story of Track-Type Tractors from 1713 to 1904," 14-page typescript, 1967. F. Hal Higgins Collection.
2 Edward Eyre and William Spotswood, "Drawings of Great Britain Patents. Number 5260," (London, England, 1856), Compiled by Dewey, Strong, Townsend and Loftus, San Francisco, 1923. The Haggin Museum.
3 Clark C. Spence, "God Speed the Plow: The Coming of Steam Cultivation to Great Britain," (Urbana, Illinois, 1960), pages 15-16.
4 "Harper's Weekly," September 12, 1857, page 589.
5 *Op. Cit.*, "Drawings of Great Britain Patents," It is interesting to note that the Warren Miller steam crawler tractor included all the basic features of modern track-type tractors. They were:
1. Main rectangular frame to carry the steam engine and boiler.
2. A supplemental frame parallel with the main frame for holding the two driving sprocket wheels.
3. A set of sprockets drove an endless chain track on each side of the engine.
4. The tracks consisting of links, flexibly joined to each other by link pins, which engaged the sprocket wheels. The treads were equipped with removable shoes.
5. Rollers supporting the weight of the engine ran on the inner surface of the tracks.

6 "United States Department of Agriculture Annual Reports" 1869, Washington, D. C.
7 Lore A. Rogers and Caleb W. Scribner, "The Lombard Log Haulers,"*Live Steam*, (Cadillac, Michigan), October 1975, pages 8-10.
8 The Holt Manufacturing Company vs. C. L. Best Gas Traction Company, Number 167. United States District Court for Northern California. "Testimony of Alvin O. Lombard," December 9, 1915. Federal Records Center, San Bruno, California.
9 *Ibid.*, page 145.
10 *Ibid.*, pages 3-16, 46.
11 *Ibid.*, pages 3-16.
12 Lore A. Rogers and Caleb W. Scribner, "Don't Jump for Your Life, Sam,"*Live Steam*, September 1975, pages 6-8.
13 "Fifty Years of Tracks," (Peoria, Illinois, 1954), page 12.
14 C. Parker Holt, "Early History of the Holt Manufacturing Company," 22-page typescript, 1935. The Haggin Museum.
15 George L. Dickenson to Benjamin Holt, September 30, 1903. Pliny Holt Papers. The Haggin Museum.
16 *Op. Cit.*, "Fifty Years of Tracks", page 12.
17 Interview with Herb Shepard, Stockton, California, by Raymond Hillman, August 7, 1975.
18 George L. Dickenson, "Diary 1893-1909," Entries on March 29, 1905 and April 4, 1905.
19 "Farm Implement News" (Chicago, Illinois) May 18, 1905, page 1.
20 *Op. Cit.*, C. Parker Holt, page 15.
21 F. Hal Higgins, "I Was the Boy Who,"*Iron-Men Album*, September-October, 1953, page 12. Higgins had interviewed Paul E. Weston in Stockton on June 20, 1953 when he copied the original order for this traction engine.

CHAPTER 8

1 "American Thresherman and Farm Power," July 1916, page 28.
2 Margaret E. Robinson, "Daniel Best Biography,"*Engineers & Engines*(Joliet, Illinois), June 1969, page 206.
3 *Ibid.*, page 5.
4 George L. Dickenson, "Diary 1893-1909," October 6, 1906.
5 *Ibid.*, December 4, 1906.
6 "Tractors & Motors Built at Stockton from 1908 to 1917," Business Records, Holt Manufacturing Company. Pliny Holt Papers. The Haggin Museum.
7 Production Record Book, 1894-1912, Number LR-67-7332, 354. Business Records, Holt Manufacturing Company. Pliny Holt Papers. The Haggin Museum.
8 Lynwood Bryant, "The Invention of the Internal Combustion Engine,"*Publications in the Humanities,* Number 81. Massachusetts Institute of Technology, Cambridge, Massachusetts, 1966, pages 184-200.
9 Reynold M. Wik, "The Invention of the Farm Gasoline Tractor," *Actes du XI Congress International d'Histoire des Sciences* (Warsaw, Poland), 1966, vol. 4, pages 205-210.
10 A. C. Seyforth, "Treatise on the History of the Tractor," 24-page typescript. McCormick Library, Chicago, Illinois, 1923, page 5.
11 C. R. Massinger, "How a New Word was Added to the English Language," Pamphlet, (Chicago, Illinois: Oliver Tractor Company, 1935), page 5.
12 Transactions of the American Society of Agricultural Engineers, (St. Paul, Minnesota, 1980), 3:21; George L. Dickenson to Murray M. Baker, Peoria, Illinois, in Murray M. Baker Correspondence Files given to R. M. Wik in 1947 by Baker.
13 "Horseless Age," June 23, 1915, page 835. Anticipating the trend to smaller tractors, the Bull Company of Minneapolis built a small tractor in 1913 which weighed only 4,650 pounds. Other small tractors were made by Wallis, Avery, Bates, and Happy Farmer. In 1915, Henry Ford announced his entrance into the small tractor field.
14 "Holt Manufacturing Company Catalogue," 1912. The Holt Caterpillar tractor of 1912 had a one-piece crank case which held a four-throw crankshaft with a nine-inch bearing at the at the rear of the motor which supported a 420-pound flywheel. Lubrication was achieved by both a splash and gear pump system. The motor cylinders were cast singly to accommodate pistons with a 6-1/2-inch bore and an 8-inch stroke, producing 45 horsepower at 550 r.p.m. The clutch operated off an open flywheel, but the transmission gears were enclosed in an oil bath. The differential consisted of a tail shaft with a pinion which engaged a large 106-tooth steel pinion mounted on a counter shaft which in turn carried the power to two drive chain sprockets. The drive shafts were divided in order to let one track move independently of the other, thus providing a steering mechanism which could turn the tractor in a 30-foot radius. When pulling a load on the drawbar, the tractor transmitted power directly to the driving sprockets, thus avoiding the usual three-gear reduction in the transmission. The Holts claimed this eliminated a 7 percent loss in power.
15 *Ibid.*, page 16. The interchangeable steel shoes which gripped the ground were made of pressed steel. They came in three sizes: 16 inches, 24 inches, and 36 inches wide.
16 Carl Crow, "Uncle Ben and the Caterpillar,"*Sunset*, January 1917. Reprinted in Paul C. Johnson, ed., "The Early Sunset Magazine, 1898-1928" San Francisco, 1974.
17 George L. Dickenson to Pliny E. Holt, Minneapolis, Minnesota, March 3, 1909; Ben C. Holt to Pliny E. Holt, Minneapolis, Minnesota, November 12, 1909. Both in Pliny E. Holt Papers.
18 R. S. Springer to Pliny E. Holt, Minneapolis, Minnesota, July 9, 1909. Pliny E. Holt Papers.
19 R. S. Springer, "Early History of the Holt Manufacturing Company," 10-page manuscript, 1929. The Haggin Museum. Springer states that between 1908 and 1921, from 12 to 15 different types of small tractors were built at various Holt plants.
20 *Ibid.*, page 9.
21 "Stockton Daily Evening Record," March 8, 1904, page 8.
22 *Ibid.*, March 16, 1904, page 6.
23 *Ibid.*, July 6, 1904, page 3.
24 Dickenson, "Diary,"March 8, 1906.
25 "Financial Statement of the Aurora Engine Company,"October 9, 1910, page 3. Pliny E. Holt to George L. Dickenson, Stockton, California, September 27, 1909, Pliny E. Holt Papers.
26 Interview with Harold J. Baker, Sunnyvale, California, by Raymond Hillman and Reynold Wik, June 6, 1975.
27 Interview with Walter Gilgert, Stockton, California, by Raymond Hillman and Reynold Wik, June 10, 1975.

CHAPTER 9

1 Production Book, pages 20-21. Business Records, Holt Manufacturing Company, Stockton, California. The production book listed the name of the party purchasing an engine, the date of sale, the shipping date, the address of the purchaser, and a brief description of the machine.
2 "Stockton Daily Evening Record," September 25, 1908, page 1.
3 *Ibid.*, page 1.
4 Dickenson, "Diary,"November 1, 1908.
5 R. S. Springer to E. H. Wedekind, Superintendent, The Groom Nevada Lead Mines Company, Goldfield, Nevada, February 19, 1909, Pliny E. Holt Papers.
6 Russell S. Springer to George L. Dickenson, Stockton, California, March 23, 1909. Pliny E. Holt Papers.
7 George L. Dickenson to Pliny E. Holt, Minneapolis, Minnesota, June 9, 1909. Pliny E. Holt Papers.
8 Higgins, "I Was the Boy Who,"page 13.
9 "Caterpillar Traction Engine Mile Tonnage Costs,"Operations from Siding S. Jawbone Division, File 2. Pliny E. Holt Papers. Operating costs included: driver, $4.17 per day; tender, $2.25 per day; depreciation, $8.00 per day; oil, 3.1 cents per gallon. For 26 days of operation, the cost of oil for fuel was $68.20, bringing the cost for 655 miles of freight hauling to 67.5 cents per ton mile.
10 R. S. Springer to Paul E. Weston, Mojave, California, February 13, 1909. Pliny E. Holt Papers.
11 R. S. Springer to the Tractor Division, Holt Manufacturing Company, Stockton, California, April 2, 1909. Pliny E. Holt Papers.
12 R. S. Springer to Pliny E. Holt, Minneapolis, Minnesota, April 23, 1909. Pliny E. Holt Papers.

13 R. S. Springer to J. B. Lippincott, Assistant Engineer, Los Angeles Aqueduct, Los Angeles, California, April 10, 1909. Pliny E. Holt Papers.
14 "Los Angeles Aqueduct Engineer's Final Report,"Department of Public Service, 1916, 3-page typescript. F. Hal Higgins Collection.
15 Interview with Paul E. Weston, Stockton, California, by F. Hal Higgins, June 10, 1952. F. Hal Higgins Collection.
16 Barton W. Currie, "The Tractor and Its Influence Upon Agricultural Implement Industry," (Philadelphia, 1916), pages 137-144.
17 R. S. Springer, Stockton, California to Pliny E. Holt, Minneapolis, April 1, 1909. Pliny E. Holt Papers.
18 "Stockton Daily Evening Record," September 25, 1908, page 1.

CHAPTER 10

1 Letter from Feeney and Feeney, Chartered Patent Agents. London, England, to Holt Manufacturing Company, January 19, 1920. Business Records, Holt Manufacturing Company. The Haggin Museum.
2 Margaret E. Robinson, "From Lumber and Grain to Tracks," 49-page typescript, May 15, 1963, page 11. F. Hal Higgins Collection.
3 Lynn W. Ellis and Edward A. Rumley, "Power and the Plow," (New York, 1911), page 3.
4 "King Alfonso Viewed A Stockton Wonder,"*Stockton Daily-Evening Record,* June 3, 1904, page 1.
5 "Detailed Report of Russian Tractor Contest," 20-page typescript, 1916, pages 6-12. F. Hal Higgins Collection.
6 "Exhibitors Weekly Bulletin," Panama-Pacific International Exposition, (San Francisco, California), vol. 6, no. 6, August 21, 1915, page 4.
7 Reynold M. Wik, "Nebraska Tractor Shows, 1913-1919, and the Beginning of Power Farming",*Nebraska History,* Summer 1983, pages 193-208.
8 "Breeders Gazette," August 24, 1904, page 292.
9 Fred Y. Portal, Fort Worth, Texas to Henry Ford, Dearborn, Michigan, June 2, 1909. Henry Ford Business Correspondence, Accession 94, Box 93. Ford Motor Company Office Building, Dearborn, Michigan.
10 "Northwest Homestead,"(St. Paul, Minnesota), December 9, 1911, page 4.
11 "The Gas Review,"(Madison, Wisconsin), October 1908, page 17.
12 "Kansas Farmer," September 12, 1914, page 5.
13 "Caterpillar Times,"(Peoria, Illinois), February 1915, page 15.
14 *Ibid.,* page 16.

CHAPTER 11

1 "Stockton Daily Record," December 24, 1903, page 8. Also, January 8, 1906, page 8.
2 *Ibid.,* May 24, 1906, page 1.
3 William Hurst and Lillian Church, "Power and Machinery in Agriculture," United States Department of Agriculture Miscellaneous Bulletin. No. 157, Washington, D. C., 1933, page 7.
4 Fred A. Wirt, Advertising Manager, J. I. Case Company, Racine, Wisconsin, to Reynold M. Wik, Minneapolis, Minnesota, November 11, 1947.
5 Murray M. Baker, "Establishing Peoria's Largest Industry," 17-page typescript, page 4. F. Hal Higgins Collection.
6 "Eleventh Annual Report of the Department of Agriculture of of the Province of Saskatchewan,"(Regina, Canada), 1916, page 139. Also, "Minneapolis Journal," July 23, 1903, page 3.
7 George L. Dickenson, Stockton, California to Pliny E. Holt, Minneapolis, Minnesota, March 19, 1909.

Unless otherwise described, all footnoted correspondence hereafter in this chapter is from the Pliny E. Holt Papers, Haggin Museum, Stockton, California.

8 "December Trial Balance, Norhtern Holt Company," C. J. Gotshall to Pliny Holt, October 7, 1909. The balance sheet showed the assets to be $1,091,448.
9 Ben C. Holt, Walla Walla, Washington, to Pliny Holt, Minneapolis, Minnesota, April 14, 1909.
10 "The Peoria Star," October 26, 1964. See also, F. Hal Higgins, "M. M. Baker Looks Back at 80,"*Iron-Men Album* November-December 1952, page 3. Baker was born near Alton, Illinois in 1872 and died in 1964 at age 92.
11 Legal Papers, Attorney S. M. Dabney Files, Caterpillar Tractor Co. Baker, "Establishing Peoria's Largest Industry," page 4.
12 Pliny Holt to Murray M. Baker, Peoria, Illinois, May 13, 1909.
13 Pliny Holt to Ben C. Holt, Peoria, Illinois, July 1, 1909.
14 M. M. Baker to Pliny Holt, Minneapolis, Minnesota, July 15, 1909; George L. Dickenson to Pliny Holt, Minneapolis, Minnesota, July 9, 1909; Ben C. Holt to Pliny Holt, Peoria, Illinois, July 17, 1909.
15 Baker, "Establishing Peoria's Largest Industry,"page 10.
16 Pliny Holt to Ben C. Holt, Walla Walla, Washington, July 28, 1909.
17 Ben C. Holt, Walla Walla, Washington, to William Stone, Chicago, Illinois, August 2, 1909.
18 Baker, "Establishing Peoria's Largest Industry," page 13.
19 Pliny Holt to M. M. Baker, Peoria, Illinois, September 29, 1909.
20 Pliny Holt to C. Parker Holt, Balboa Building, San Francisco, California, September 30, 1909.
21 C. Parker Holt to Pliny Holt, Minneapolis, Minnesota, October 9, 1909.
22 M. M. Baker to Pliny Holt, Minneapolis, Minnesota, in Baker "Establishing Peoria's Largest Industry," page 16; Baker to Frank Harrison Holt, Potomac, Maryland, January 5, 1955.
23 C. Parker Holt to Pliny Holt, Minneapolis, Minnesota, December 6, 1909.
24 Legal Papers, Attorney S. M. Dabney Files, Caterpillar Tractor Co.
25 Ben C. Holt to Pliny Holt, Minneapolis, Minnesota, January 2, 1910. See also, G. L. Dickenson to Ben C. Holt, Walla Walla, Washington, December 27, 1909.

CHAPTER 12

1 Interview with Murray M. Baker, Peoria, Illinois by Reynold M. Wik, July 6, 1947.
2 Pliny E. Holt, Peoria, Illinois, to Ben C. Holt, Walla Walla, Washington, June 20, 1910. All footnoted correspondence in this chapter is from the Pliny E. Holt Papers, The Haggin Museum, Stockton, California.
3 Murray M. Baker, Peoria, Illinois to Frank Harrison Holt, Potomac, Maryland, January 5, 1955.
4 Ben C. Holt, Walla Walla, Washington, to Pliny E. Holt, Minneapolis, Minnesota, November 4, 1909.

5 C. Parker Holt, Stockton, California, to Ben C. Holt, Walla Walla, Washington, October 22, 1910.
6 Pliny E. Holt, Peoria, Illinois, to Ben C. Holt, Walla Walla, Washington, July 6, 1910.
7 H. L. Freeman, Regina, Saskatchewan to the Holt Caterpillar Company, Peoria, Illinois, September 30, 1910.
8 John Corrigan, Regina, Saskatchewan, to Pliny E. Holt, Peoria, Illinois, October 29, 1910.
9 "Farm Implement News," (Chicago, Illinois), April 1899, page 14.
10 F. Hal Higgins, "Super Salesmen in the Tractor Business," 14-page typescript, June 2, 1947. F. Hal Higgins Collection.
11 Ben C. Holt, Walla Walla, Wasington, to Pliny E. Holt, Peoria, Illinois, July 7, 1910.
12 F. B. Franklin, Bloomington, Illinois to Murray M. Baker, Peoria, Illinois, January 2, 1908. M. M. Baker & Company Business Correspondence. This file is now in the possession of R. M. Wik.
13 Ben C. Holt, Walla Walla, Washington, to Pliny E. Holt, Peoria Illinois, August 26, 1909.
14 Pliny E. Holt, Peoria, Illinois, to Ben C. Holt, Walla Walla, Washington, October 3, 1910.
15 Pliny E. Holt, Peoria, Illinois to Ben C. Holt, Walla Walla, Washington, November 6, 1910.
16 Ben C. Holt, Walla Walla, Washington to Pliny E. Holt, Peoria, Illinois, April 19, 1910.

CHAPTER 13

1 "Nine More Holt Caterpillar Gasoline Traction Engines for South America," *Holt Bulletin* (Stockton, California), April 1, 1911, page 1.
2 F. Hal Higgins, "Death Comes to C. Parker Holt,"*Farm Implement,* September 8, 1938, pages 24-25.
3 "Record of Traction Engines, Wagons, Freighting and Steam Harvesting Outfits," 1911, pages 126-193. Business Records, Holt Manufacturing Company, Stockton, California. Pliny E. Holt Papers. The Haggin Museum.
4 Charles L. Neumiller, "Tractors and Motors Built at Stockton from 1908 to 1917," Business Records. Holt Manufacturing Company. Also, "Amortization Claim Report to the Commissioner of Internal Revenue" June 16, 1923. Business Records, Holt Manufacturing Company. Pliny E. Holt Papers. The Haggin Museum.
5 C. L. Best, "Report and Financial Statement to the Board of Directors of the Best Manufacturing Company,"January 15, 1910. Pliny E. Holt Papers. The Haggin Museum.
6 "Stockton Daily Independent," December 26, 1916, page 4.
7 Interview with William K. Holt, Hillsborough, California by Reynold M. Wik, November 10, 1975.
8 "Account Journal Entries," and "Original Contractual Agreements between Benjamin Holt, Pliny E. Holt, C. Parker Holt, Ben C. Holt, and Russell S. Springer," June 20, 1913. Business Records, Holt Manufacturing Company, Stockton, California. The Haggin Museum.
9 "New York Evening Journal," July 1, 1913, page 1.
10 *Ibid.,* September 1, 1913, page 1.
11 "Wallace's Farmer,"(Des Moines, Iowa), February 11, 1916, page 6.
12 Raymond W. Hillman, "Holt 75 Caterpillar Track-Type Tractor," *Holt Memorial Hall,* June 1976, page 5. The Haggin Museum. See also, "The Weekly Oregonian,"(Portland, Oregon), December 19, 1912.

CHAPTER 14

1 Rollin W. Hutchinson, Jr., "The Motor Versus the Mule in Uncle Sam's War Department,"*The American Review of Reviews,* July 1913, pages 59-64, 63.
2 Ben C. Holt to Senator Miles Poindexter, November 26, 1913, Department of War, Box 287, R. G. Number 165; Stockm 16w-3, Row 4, Compartment 30, Shelf F., Card number: 181942, W.C.D. 8479-8, National Archives.
3 Ben C. Holt to Board of Ordnance and Fortifications, December 17, 1913; Captain of the Corps of Engineers, Board of Ordnance and Fortifications, War Department to Ben C. Holt, New York, December 20, 1913; Miles Poindexter to Major General Leonard Wood, Board of Ordnance and Fortifications, December 23, 1913; Major General Leonard Wood, Chief of Staff, War Department, to Senator Miles Poindexter, United States Senate, Washington, D. C., December 30, 1913, National Archives.
4 Ben C. Holt to Board of Ordnance and Fortifications, War Department, May 11, 1914, National Archives.
5 Murray M. Baker, "The Holt Caterpillar Company and World War I," 9-page manuscript, November 8, 1919. F. Hal Higgins Collection.
6 *Ibid.,* page 5.
7 "New York Times," September 20, 1916.
8 Arthur S. Link, "American Epoch," (New York, 1955), pages 165-171.
9 Major General John J. Pershing, "Motor Truck Transportation in Mexico,"*Texas Motorist* April 1917, pages 5-7.
10 Byron DeHaan, "The Holt Tractor: The Beginning of Warfare's First Tanks,"*News and Views,* May 1953, pages 12-13. See also, "Record of Traction Engines, Wagons, Freighting and Steam Harvester's Sales," Business Records, Holt Manufacturing Company.
11 "Fifty Years on Tracks," page 24.
12 DeHaan, "Holt Tractor," page 14. Interview with William K. Holt, November 8, 1977.
13 B. T. White, "Tanks and Other Armored Fighting Vehicles, 1900-1918," (New York, 1970), page 9.
14 C. Parker Holt, "Early Developments of the Holt Manufacturing Company" page 19.
15 "Stockton Daily Evening Record," September 19, 1916;*The San Francisco Bulletin,* September 19, 1916.
16 "To the Front on a Caterpillar,"*The Auto Car* November 13, 1915, pages 607-608.
17 "Scientific American," Supplement No. 2124, September 16, 1916, page 184; Kenneth Macksay and John H. Batchelor, "Tank," (New York, 1971), page 27; White "Tanks and Other Vehicles," page 5.
18 Major E. D. Swinton, Manuscript of speech given in Stockton on April 22, 1918. The Haggin Museum. Macksay and Batchelor, "Tank," page 24.
19 John H. Batchelor, "The Tank Story."
20 "New York Herald," October 28, 1917, page 1.
21 Op. Cit., manuscript of speech, page 8.
22 Ralph E. Jones, George H. Rarey, Robert J. Icks, "The Fighting Tanks," (Washington, D. C., 1933), page 5.
23 "The London Times," Septemeber 16, 1916, pages 8-9, September 18, 1916, page 9, September 19, 1916, page 10,*The Scotsman* (Edinburgh, Scotland), September 18, 1916.
24 "San Francisco Daily News," October 6, 1916; "New York Evening Journal," September 21, 1916. See also "The Literary Digest," September 30, 1916, pages 815-817.

25 See, for example, "The Peoria Journal," September 20, 1916, "Washington Post," September 19, 1916, "Chicago Herald," September 19, 1916.
26 Reginald T. Townsend, "Tanks and the Hose of Death," *World's Work,* December 1916, page 195, *New York Evening Journal,* September 21, 1916.
27 "Stockton Evening Mail," October 12, 1916; "Stockton Daily Independent" October 12, 1916; "Stockton Daily Evening Record" October 12, 1916, October 13, 1916.
28 "The Stockton Daily Independent," September 13, 1916, October 8, 1916.
29 *Ibid.,* February 14, 1917.

CHAPTER 15

1 "Stockton Daily Evening Record," May 2, 1917, June 13, 1917, May 17, 1917, April 17, 1917.
2 *Ibid.,* April 21, 1917.
3 "Stockton Daily Independent," April 16, 1918, April 10, 1918.
4 *Ibid.,* April 8, 1918.
5 *Ibid.,* April 20, 1917, Februrary 4, 1918. See also, "The Daily Stockton Independent," April 21, 1917.
6 Interview with Joe Reposa, Stockton, California by Reynold M. Wik, April 16, 1976.
7 "Stockton Daily Evening Record," April 23, 1917.
8 Interview with Fred M. Ballew, Stockton, by Renold M. Wik. April 16, 1976.
9 "The Second Liberty Loan," *Bulletin No. 17,* October 8, 1917. (Published by Minute Men). The Haggin Museum.
10 "Stockton Evening Mail," June 9, 1917, page 1. See also, "Stockton Daily Evening Record," June 3, 1917; "Stockton Daily Independent," June 10, 1917.
11 Interview with Fred M. Ballew by Reynold M. Wik April 16, 1976. "The Tank," (Stockton: Holt Manufacturing Company), vol. I, no. 5, July 1919, pages 7-9.
12 Interview with Jack Ross, San Andreas, California by Reynold M. Wik and Raymond Hillman, April 15, 1976.
13 "Stockton Evening Mail," June 19, 1917.
14 "Stockton Labor Review," June 13, 1918, June 20, 1918.
15 "Stockton Daily Evening Record," May 16, 1917, May 17, 1917, July 8, 1918.
16 *Ibid.,* April 9, 1917, March 30, 1918.
17 Murray M. Baker, "A Brief History of the Claim to the Commissioner of Internal Revenue," 16 pages, June 16, 1923, Business Records, Holt Manufacturing Company.
18 *Ibid.,* pages 5-6, "Peoria Star," September 20, 1917.
19 Murray M. Baker, "History of the Holt Caterpillar Company in Peoria," 9-page manuscript, January 19, 1919. F. Hal Higgins Collection.
20 "Specifications for Motor Transport, 1916-1920" Ordnance Department, Office of the Chief of Ordnance, United States Army, Circular Advertisements for Sealed Bids for Gasoline Tractors, Box 1, Entry 754, R. G. 156, February 8, 1916, March 1, 1916, and August 31, 1916, War Department Archives, National Archives.
21 "The Peoria Star," December 10, 1916.
22 "Op. Cit.," Baker, "Brief History of the Claim to the Commissioner of Internal Revenue," page 8.
23 Henry G. Sharpe, Brigadier-General, Quartermaster Corps, Ordnance Department, Entry 754, Box 2, June 8, 1916, National Archives.
24 H. W. Alden, Lieutenant Colonel, Ordnance Department, Tank, Tractor and Trailer Division, "Report on the Data on 10, 15, and 20-ton Tractors," February 11, 1919. National Archives.
25 "Ordnance Department Contracts for Holt Manufacturing Company, Peoria, Illinois," compiled by the Army Inspector of Ordnance, Manufacturing, Service, Tank, Tractor and Trailer Division, O.M.T. Decimal Correspondence Files, 1919-1920, Box 50, No. 451,22. National Archives.
26 Daniel Gilmore to Pliny E. Holt, Washington, D. C., November 14, 1917.
27 "Number of Holt Caterpillar Tractors Built," Business Correspondence, Holt Manufacturing Company.
28 Interview with William K. Holt by Reynold Wik, Hillsborough, California, April 3, 1976.
29 Murray M. Baker to Colonel George W. Burr, Ordnance Department, Rock Island Arsenal, September 27, 1916, Business Correspondence, Holt Manufacturing Company.
30 "San Francisco Bulletin," January 10, 1921; "Caterpillar Times," May 1918, page 5. See also, "Stockton Daily Independent," January 10, 1912.
31 "Op. Cit.," Swinton, speech manuscript, Stockton, California April 22, 1918.
32 Daniel Gilmore to Pliny Holt, Washington, D. C., November 14, 1918, Business Correspondence, Holt Manufacturing Company.
33 "Consolidated Balance Sheet. Profits & Loss Statements," Financial Records, January 1918, Business Correspondence, Holt Manufacturing Company; "San Francisco Chronicle," March 27, 1919. Figures given by Thomas Baxter, General Manager of the Holt Manufacturing Company in 1919.
34 "Consolidated Balance Sheet," January 1918. See also, "San Francisco Chronicle," March 27, 1919.

CHAPTER 16

1 Daniel N. Gilmore to Pliny E. Holt, Washington, D. C., November 14, 1918, Business Records, Holt Manufacturing Company.
2 Murray M. Baker to Procurement Division, Ordnance Department, July 26, 1918, Decimal Correspondence Files, 1919-1920, Number 451-22, Box 50.
3 Samuel W. Blythe to Pliny E. Holt, Washington, D. C., December 11, 1919, in Pliny E. Holt Papers; Colonel L. B. Moony, Ordnance Department, War Department, to Army Inspector, Holt Manufacturing Company, Peoria, Illinois, June 23, 1919. Number 451-22-777. National Archives.
4 Murray M. Baker, "Brief History of the Claim" Charles L. Neumiller, "Amortization Claim," Report to the Commissioner of Internal Revenue, June 16, 1923, Exhibit VIII, 1914 to 1923, Business Records, Holt Manufacturing Company.
5 "Charts of Total Sales: Tractor Sales, Harvester Sales, and Spare Parts Sales." Pliny E. Holt Papers.
6 "Financial Records, 1908 to 1923," and "Surplus and Net Earnings for the Years 1892 to 1923 Inclusive." Pliny E. Holt Papers. James H. Shideler, "The Development of the Parity Price Formula for Agriculture, 1919-1923," *Agricultural History,* July 1953, page 81; "Farm Implements" Minneapolis, Minnesota, December 30, 1922, page 4.

7 W. F. Jolly, London, England, to M. M. Baker, Peoria, Illinois, August 15, 1922, Business Records, Holt Manufacturing Company, Murray M. Baker, "Brief History of the Claim," page 11.

8 "Denver Post," August 1, 1916, page 1. Henry Ford Business Correspondence, Ford Archives, Ford Motor Company, Dearborn, Michigan. See also, Reynold M. Wik, "Henry Ford's Tractors and American Agriculture,"*Agricultural History*,April 1964, pages 79-80.

9 W. H. Yount, Engineering Department, Ordnance Department, "Report of Drawing Release and Change Notices for 1923," January 3, 1924. Pliny E. Holt Papers.

10 Ray M. Hart, Office of the Secretary, Department of Commerce, Washington, D. C., to M. M. Baker, Peoria, Illinois, December 28, 1922. Pliny E. Holt Papers; Secretary of War, War Department Files, No. 451, 22 1149, Memorandum, May 22, 1923. National Archives.

11 Archer P. Whallon, "There Were Giants in Those Days,"*The Farm Quarterly* Cincinnati, Ohio, Spring 1947, pages 22-26.

12 Captain T. A. Collins, Ordnance Department, to Holt Manufacturing Company, Peoria, Illinois, January 10, 1919, F. L. Peterson, Service Engineer, to H. C. Buffington, Chief Engineer, Holt Manufacturing Company, Peoria, Illinois, October 30, 1919. War Department Files, Federal Records Center, Suitland, Maryland.

13 Interview with Pliny E. Holt, Potomac, Maryland, by Reynold M. Wik, October 20, 1975.

14 Interview with Ellen Drake Holt, Stockton, California, by Reynold M. Wik, April 16, 1976.

15 Interview with William K. Holt, April 1976.

16 "Consolidated Balance Sheets; Profit & Loss Statements," Holt Manufacturing Company, January 1918 and January 1924. Pliny Holt Papers.

17 Thomas F. Baxter to Benjamin Holt, Stockton, California, March 11, 1913.

18 Pliny E. Holt to Thomas F. Baxter, Bond & Goodwin, Boston, Massachusetts, December 8, 1915.

19 Interview with William K. Holt, May 1976.

20 George D. Babcock to Thomas F. Baxter, Stockton, California, October 6, 1923.

21 H. B. Baker to M. M. Baker, Peoria, Illinois, April 8, 1924. See also, "Advertising of the Holt Manufacturing Company," LR 67-7797-381-417. The Haggin Museum.

22 "Pacific Rural Press," January 24, 1919.

23 "Analysis of Sales. Block 1-16," 18 pages. Pliny E. Holt Papers.

24 "Fifty Years on Tracks," page 37.

25 Howard C. Wilson to the Holt Manufacturing Company, Peoria, Illinois, February 9, 1925.

26 H. G. Pine to Holt Manufacturing Company, Peoria, Illinois, April 26, 1924; Charles M. Ingalls, Rome, Georgia, to M. M. Baker, Peoria, Illinois, June 6, 1923; John L. Swenton, Daytona, Florida, to M. M. Baker, Peoria, Illinois, May 12, 1922.

27 Murray M. Baker to Thomas F. Baxter, Stockton, California, February 12, 1923. Pliny E. Holt Papers.

28 "Fifty Years on Tracks," page 28.

29 Babcock to Baxter, October 6, 1923; "Memoranda of Minutes of Meeting held January 30, 1924." Pliny E. Holt Papers.

30 F. W. Tarr, Assistant Auditor, Holt Manufacturing Company, "Financial Statements and Inventory," June 30, 1924.

31 *Ibid.*, pages 2-3.

32 "Prospectus of the Holt Manufacturing Company." Pliny E. Holt Papers.

33 "Peoria Journal Transcript," March 3, 1925. See also, "Peoria Labor Gazette," March 6, 1925.

34 "Peoria Star," March 2, 1925.

35 "Prospectus of the Holt Manufacturing Company," pages 1-4. Pliny E. Holt Papers.

36 Interview with Harry Holt, by Reynold M. Wik, Stockton, December 13, 1977; "Fifty Years on Tracks," page 28.

CHAPTER 17

Most of the court cases which involved the Holt Manufacturing Company were tried in the Circuit Court of the United States for the Ninth Circuit for Northern California. These court records are now housed in the Federal Archives and Records Center in San Bruno, California.

1 Case No. 14628, in the Circuit Court of the United States for the Ninth Judicial District for the Northern District of California, The Best Manufacturing Company vs. The Holt Manufacturing Company.

2 *Ibid.*, Case No. 1608, No. 13583.

3 *Ibid.*, page 11.

4 *Ibid.*, pages 148-149.

5 "Transcript of Record," Case No. 1608, filed May 15, 1908, No. 13583, page 4.

6 United States Circuit Court of Appeals for the Ninth Circuit, "Transcript of Record," vol. II, The Holt Manufacturing Company (Plaintiff in Error) vs. The Best Manufacturing Company (Defendant in Error), No. 1608, "Petition for Writ of Error," by I. M. Kalloch and F. St. J. Fox, pages 845-863.

7 *Ibid.*, page 559.

8 Ben C. Holt to C. Parker Holt, San Francisco, California, August 3, 1909. Pliny E. Holt Papers.

9 Charles E. Townsend, "History of the C. L. Best Gas Traction Company," March 26, 1917, manuscript, pages 2-4. Pliny E. Holt Papers.

10 Townsend, "History of C. L. Best Company," pages 6-12.

11 *Ibid.*, pages 7, 8-14.

12 Malloch & Fox to The Holt Manufacturing Company, August 27, 1906.

13 Pennie, Goldsborough & O'Neill, Washington, D. C., to Pliny E. Holt, Stockton, California, January 14, 1909.

14 *bid.*, pages 12, 26.

15 *Ibid.*, pages 13, 25-26.

16 P. E. Ehrenfeldt, Secretary, "Memorandum and Exhibits Submitted to the Treasury Department with Respect to Federal Tax Matters for the Years 1917 to 1921 Inclusive," July 29, 1925, Exhibit C, Business Records, The Holt Manufacturing Company.

17 E. L. Thurston to C. L. Best Gas Traction Company, San Leandro, California, March 11, 1907. Pliny E. Holt Papers.

18 Holt Manufacturing Company vs. The C. L. Best Gas Traction Company, No. 167, in Equity, "Transcript of Testimony," December 9, 1915, pages 123-126; Henry C. Montgomery, "Lombard Holt and Best Put Tracks to Work." Notes from F. Hal Higgins Collection.

19 Transcript of Testimony, No. 167 in Equity, December 9, 1915, pages 132, 134.

20 Depositions of C. L. Tolles et al, on behalf of C. L. Best Gas Traction Company, May 29, 1916, Case No. 167 in Equity, R. C. No. 21, Box 63, page 46. See also Deposition from Charles H. Richardson, March 6, 1916, page 10.

21 *Ibid.,* Deposition from C. L. Tolles, "Oakland Tribune," May 24, 1916, page 176. See also, "Waterville Morning Star," March 5, 1918, page 3.

22 The Holt Manufacturing Company vs. The C. L. Best Gas Traction Company, No. 167 in Equity, Final Decree dated December 30, 1918, pages 256-257.

23 There are no exact figures indicating the total costs of litigation with the Best Company. The costs of litigation to both the Bests and Holts from 1904 to 1918 may have ranged from $1 million to $2 million.

24 Montgomery, "Lombard, Holt and Best," page 4.

25 P. E. Ehrenfeldt, Secretary, Holt Manufacturing Company, "Memorandum and Exhibits Submitted to the Treasury Department with Respect to Federal Tax Matters for the Years 1917-1921," July 29, 1925, Exhibit C, Business Records, Holt Manufacturing Company, pages 7-9.

26 Charles M. Barkeley, "Legal Report,"April 8, 1916, pages 29-30, 64-72, 82, 86-89.

CHAPTER 18

1 "Caterpillar Annual Report," (Caterpillar Tractor Co. Peoria, Illinois, 1974), pages 28-29.

2 R. C. Force, San Leandro, California to Major General C. C. Williams, Chief of Ordnance, War Department, May 6, 1926. Reply, C. C. Williams to R. C. Force, May 14, 1926. Business Records, Caterpillar Tractor Co., Peoria, Illinois.

3 Robert G. LeTourneau, "Mover of Men and Mountains," (Chicago, Illinois, 1960), pages 96-97.

4 *Ibid.,* page 215.

5 Kenneth Galbraith, "The Great Crash," (New York, 1955), page 182.

6 James Hedges, Grass Valley, California to Henry Ford, Dearborn, Michigan, December 19, 1930. Ford Fair Lane Papers, Accession 380, Box 13. Ford Motor Company Archives, Dearborn, Michigan.

7 "Op. Cit.,"*Caterpillar Annual Report,* 1974,"pages 28-29.

8 Interview with William K. Holt, Hillsborough, California, April 23, 1976.

9 Remley J. Glass, "Gentlemen, the Corn Belt,"*Harpers,* July 1933, pages 200-206.

10 Reynold M. Wik, "Henry Ford and Grass-roots America,"(Ann Arbor, Michigan, 1972), page 185.

11 "Fifty Years of Tracks," (Peoria, Illinois, 1954) pages 40-42.

12 "Compton's Pictured Encyclopedia," (Chicago, Illinois, 1951), page 90.

13 Interview with Pliny E. Holt, Potomac, Maryland by Reynold M. Wik, October 22, 1976.

14 "Op. Cit."*Fifty Years of Tracks* page 43.

15 "The Saturday Evening Post," September 20, 1941, pages 16-17, 68-72. Also, August 13, 1939; July 27, 1940.

16 "The Peoria Star," March 9, 1936, page 1.

17 *Ibid.,* November 17, 1937, page 3.

18 *Ibid.,* May 19, 1935, page 6.

19 "Caterpillar Annual Report, 1942," (Peoria, Illinois), page 2.

20 F. Hal Higgins, "The Holt Manufacturing Company,"*Engineers & Engines*(Joliet, Illinois), December 1963, pages 4-6.

21 "Op. Cit.,"*Fifty Years of Tracks* page 51.

22 *Ibid.,* pages 48-51.

23 *Ibid.,* page 73.

24 "Caterpillar World,"(Peoria, Illinois), September-October, 1973; July-August, 1973; June-July, 1970.

25 "Ibid.," July-August, 1973, pages 4-8.